From BEHEMOTH to MICROSHIP

Two Platforms for Technomadic Adventure

(Fourth Edition)

By Steven K. Roberts
Nomadic Research Labs

This book is dedicated to my father, Edward H. Roberts, whose ongoing support over the years has done much to make these projects possible.

Production Staff/Fourth Edition:
Joyce Lukaczer, layout and design
Natasha Clarke, art direction and photo management
Dave Aton, copy editing
Randal Schwartz, cover photo

Published by:
Nomadic Research Labs
1313 South Hagen Road
Camano Island WA 98282
http://www.microship.com

Publishing History:

This book has evolved over the years, in parallel with the projects it describes, and this edition is by far the most comprehensive (and professionally produced) to date.

March 1994	250 copies	First edition, designed by Miria Jo and Kim Leider
July 1994	600 copies	Reprint for Nomadness subscribers
September 1995	500 copies	Second edition, designed by Faun Skyles
March 1996	500 copies	Minor edits, reprint for trade shows
June 1996	350 copies	Reprint at 1200 dpi for quality improvement
May 1999	500 copies	Third edition, designed by Natasha Clarke
July 2000	500 copies	Fourth edition, designed by Ms Deadlines (a.k.a. Joyce Lukaczer and Natasha Clarke)

Note on Project Documentation:

This book is intended to serve as a substantial overview of these technomadic projects (while capturing the flavor of the adventure), and thus includes material aimed at both technical and non-technical audiences. If you are looking for more engineering details, please consider this an introductory text and continue from here to the extensive project archives at http://www.microship.com. We also publish complete technical monographs about some of the subsystems, including schematics and software listings... please contact *wordy@microship.com* for more information.

Published July 2000
Printed in the United States of America
Fourth edition: 500 copies
Designed by Ms Deadlines
(a.k.a. Joyce Lukaczer and Natasha Clarke)

ISBN: 1-929470-00-2

Table of Contents

Chapter Five: The Microships... Io and Europa

Appendices

"Art without engineering is dreaming.
Engineering without art is calculating."

—*Steven K. Roberts*

Original Winnebiko, with 5 watts of solar panels and, at 135 lbs., light enough to lift. (Photo by reporter somewhere in Louisiana.)

Chapter One
Winnebiko & BEHEMOTH

Columbus Ohio: Inspiration for long distance travel

Way back in 1983, I was doing what most struggling young freelancers do: taking on a succession of projects, destroying old passions by turning them into businesses, and trying to make enough money to stay afloat. My lifestyle had become suburban, and as I clattered around my boring acre in an itchy haze of midwestern pollen and lawnmower smoke I wondered what had gone wrong. "Freelancing" was an illusion; I was chained to my desk and deep in debt like everyone else. Stuck.

Worse, change, evolution, and growth had begun to sound like vague counterculture concepts instead of the basic objectives of daily living.

Where had all my passions gone? One afternoon I listed them: writing, adventure, computer design, ham radio, bicycling, romance, learning, networking, publishing… each of these things had at one time or another kept me up all night in a delicious frenzy of fun and giddy intellectual growth. Yet my reality had become one of performing decreasingly interesting tasks for the sole purpose of paying bills, supporting a lifestyle I didn't like, in a house I didn't like, in a city I didn't like. I had forgotten how to play. Could it still be possible to construct a lifestyle entirely of passions, or was losing the spark a sadly inevitable part of growing up?

Combining the passions in my list and abandoning all "rational thought," the obvious solution was to simply equip a recumbent bicycle with ham radio and computer gear, establish a virtual home in the nascent online networks, and travel full time while writing and consulting for a living.

Was such a mad idea really possible?

THE GREATEST RISK

excerpt from COMPUTING ACROSS AMERICA

Everyone has at one time or another shrunk from a growth opportunity because with it comes the Unknown. The Unknown! What's out there, anyway?

Disappointment? Derision? Danger? Defeat? Death?

Those are all bad, certainly, but none of them is nearly as bad as *nothing*. None of those things can possibly be worse than the horror of Complacency that creeps like a psychic tapeworm into the mind, demanding a steady diet of the bland to let it propagate and infect those nearby. It's insidious, evil, and epidemic in America. It slithers out of TV sets; it crawls from the pages of popular media. It hides in classrooms and slips unnoticed into vulnerable young brains.

And it does all this while masquerading brilliantly as knowledge and truth. Complacency is the adulterant of passion, the assassin of curiosity, the lobotomizer of life itself.

A friend's mother once listened to me talking excitedly about my upcoming journey and shook her head in protective maternal fright. She summed it all up without even realizing it: "But honey, there are things out there... there are things out there we don't even *know* about!"

Right on.

It's possible to have growth without risk, but it is growth of the slow, vegetable kind, rarely yielding those magical breakthroughs that make you light up with understanding. It doesn't take anything quite so radical as an epic bicycle trip to do this, of course, only the courage to risk the Unknown. Any kind of Unknown – be it intellectual, geographical, cultural, scientific, physical... whatever.

Because the greatest risk of all is taking *no* risk. That's the one that can really get ya, the death worse than fate. So. Curious about something? Restless? Hungry for knowledge or change? Got a dream?

Go for it.

1983	WINNEBIKO DESIGN
1984	COMPUTING ACROSS AMERICA 10,000 MILES SOLO
1985	OHIO LAYOVER
1986	WINNEBIKO II FABRICATION MILES WITH MAGGIE 6,000 MILES
1987	
1988	SCHOOLBUS TOUR 15,000 MILES
1989	BEHEMOTH DEVELOPMENT PROJECT BIKE LAB IN SILICON VALLEY CA
1990	
1991	BEHEMOTH TOUR 1,000 MILES
1992	MOTHERSHIP TOUR 55,000 MILES
1993	MICROSHIP DEVELOPMENT PROJECT UCSD, SAN DIEGO CA
1994	
1995	
1996	MICROSHIP DEVELOPMENT PROJECT SANTA CLARA CA (APPLE)
1997	
1998	MICROSHIP DEVELOPMENT PROJECT IN PERMANENT MICROSHIP LAB ON CAMANO ISLAND WA
1999	
2000+	OPEN ENDED MICROSHIP ADVENTURES

COMPUTING ACROSS AMERICA

Six months later, to the obvious dismay of parents and girlfriend, I hit the road—having liquidated the three-bedroom ranch in suburbia and most of its contents. The bike was a custom recumbent dubbed the *Winnebiko,* the computer a little Radio Shack Model 100 laptop, the data communication link CompuServe, the power source a small five watt Solarex photovoltaic panel.

Immediately it became obvious that I had hit upon something significant. In 1983, the concept of decoupling from a fixed home without becoming a bum in the process was alien to most people, especially with the trappings of high technology allowing an active freelance lifestyle. Daily email from pay phones became routine, and soon the magazine assignments started pouring in—then a book contract for *Computing Across America.* I was interviewed in almost every town I passed through, appeared numerous times on national television, and amassed a loyal following of fellow online denizens who offered support facilities all over the country.

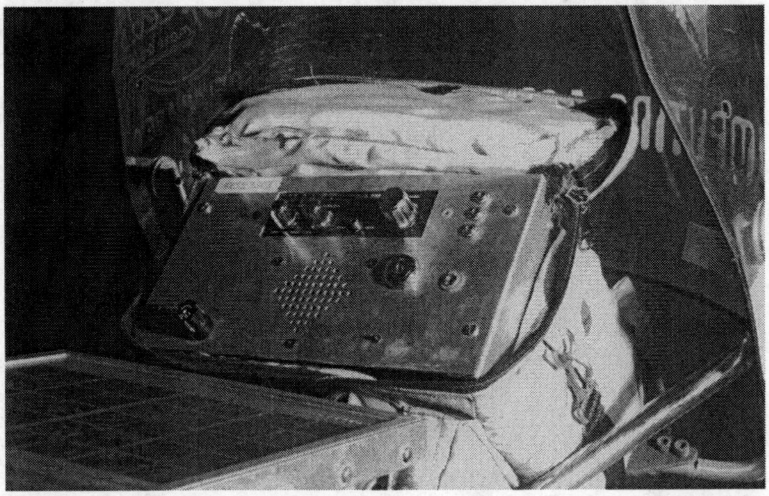

Electronics Package, containing CB, SAFT 4 amp-hour NiCd Battery, power supplies, and security system. (Photo by Katy Peden, Clearwater FL, 1984)

For 10,000 miles I wandered solo, high on the energy of beginnings and change, never really seeking publicity but becoming amazingly well known. But it wasn't me so much

Computing Across America

as the *Winnebiko*—a machine that eloquently symbolized the daring notion that people could indeed be free, follow their dreams, and break the chains that had always bound them to their desks. In small towns and big cities all across America, I saw the same faraway look, the same dawning of understanding. "Mah gawd, man, you mean to tell me you're runnin' a business from that thang? Reckon I could do somethin' like that? Always did wanna see the world..."

Some of it was the ancient lure of the road, some the glitter of new technology. ("Are you with NASA?" asked a midwest farmer one day.) But through it all I was not only championing the linked causes of freedom, risk-taking, solar power, laptops, networks, and human-powered vehicles, but I was also learning at a furious pace, and becoming deeply frustrated with the limitations of my machine.

I couldn't write while riding. I depended upon clunky wireline phone connections. I had only five watts of solar power. My radio links were tenuous at best. The evolving technology offered much more.

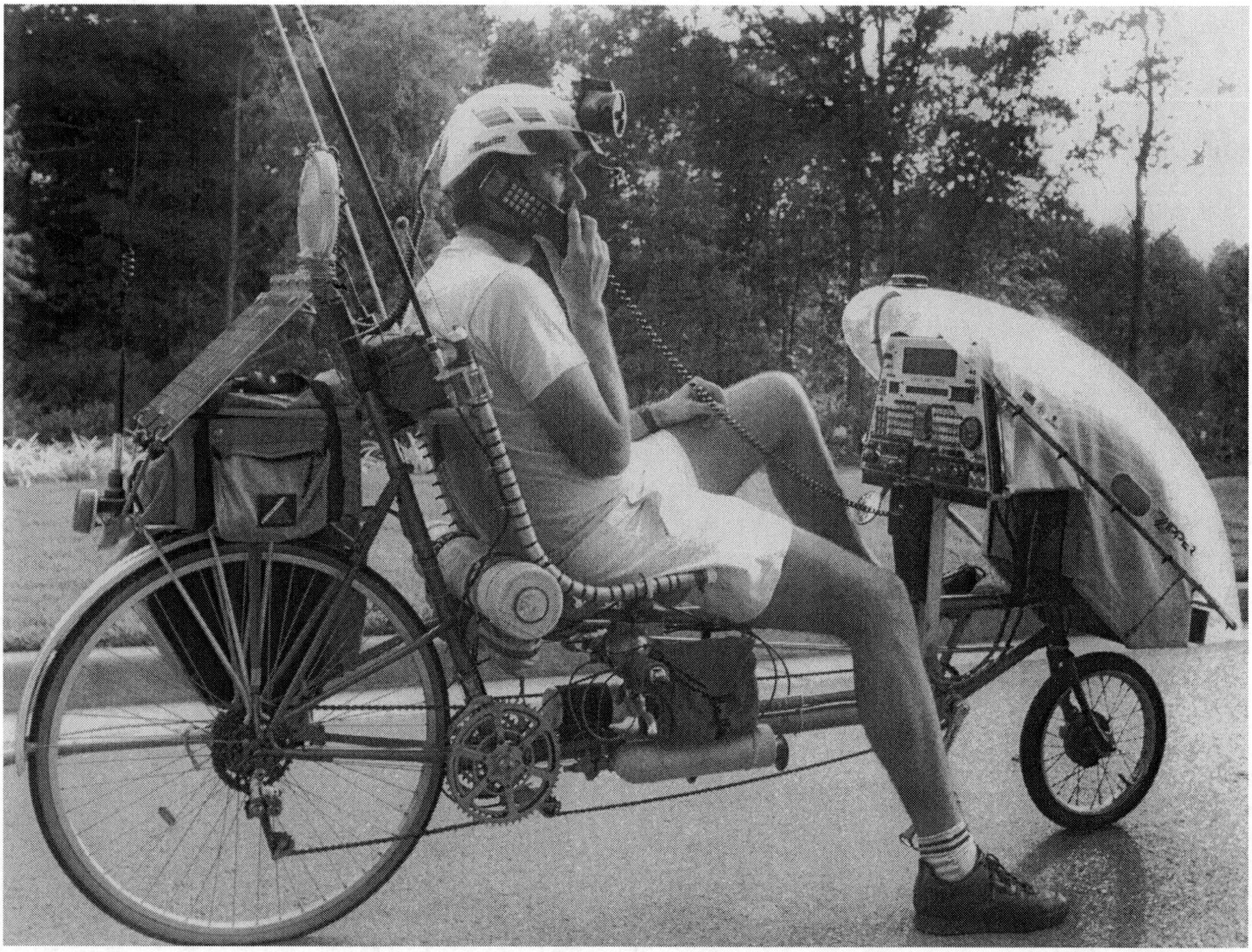

THE WINNEBIKO II AND MAGGIE

It looked like an aircraft cockpit. Bristling with switches and LEDs, the *Winnebiko II* flickered to life in the summer of 1986. The primary design objective—being able to type while riding—had evolved into a mad tangle of processors and other subsystems. The bike now had a speech synthesizer that could be triggered by security sensors or remote radio command, a packet data communication system for email via ham radio, 20 watts of solar panels, an offline HP laptop, and more. And, equally significantly, I took on a traveling companion named Maggie Victor, having grown weary of solo wandering after a year and a half of brief on-the-road relationships.

The ability to type while mobile was amazingly liberating. Every 10,000 miles is about 1,000 hours of raw pedaling time, perhaps better expressed as half of a full-time business year. Could I take advantage of it? During the first trip, I had watched brilliant tales crystallize in my head and then evaporate, lost forever in the passing breezes. Motivated by the desire to capture them, I built the new system and catapulted the adventure into a whole new phase: not only could I do word processing while riding, but I could also exchange electronic mail via packet radio.

Through ham radio and computer networking, the sense of living in a virtual neighborhood grew more and more tangible, until

> *"...a machine that eloquently symbolized the daring notion that people could indeed be free, follow their dreams, and break the chains that had always bound them to their desks."*

the road itself became merely an entertaining backdrop for a stable life in Dataspace. I moved from CompuServe to GEnie, and began exploring the embryonic Internet and many other layers of the increasingly complex network world—each a unique subculture with a different class of resources only tenuously connected to the rest.

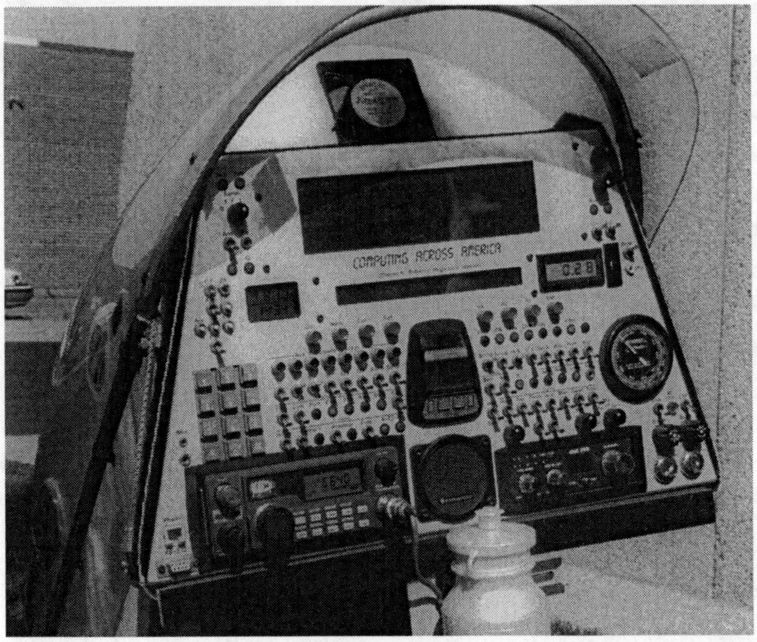

Winnebiko II console, with computer screens, DPM, clock, speed and cadence display, 2-meter ham radio, rarely-used CB, altimeter, touch-tone pad, speaker, and switches for lights, power supplies, packet, audio routing, and everything else imaginable. (Photo courtesy of GEnie Information Services, Rockville MD.)

Maggie and I pedaled 6,000 miles together, covering most of both coasts, and all along the way fired the imaginations of people who were sensing the implications of new information technologies but weren't sure what to do with them. In a small way, the bike represented the outrageous notion that very soon, it might no longer matter where your body happens to be... as long as you maintain a presence in the networks.

The Winnebiko II and Maggie

In these days of ethernet and high-speed wireless, the thought of truly decoupling via a tenuous 1200-baud link seems amusing. But the implications were staggering. Home, quite literally, became an abstract electronic concept. From a business standpoint, it no longer mattered where we were, and we traveled freely, making a living through magazine publishing and occasional consulting spinoffs, seeking modular phone jacks at every stop.

But after three years with the new bike, I could no longer ignore the advances of technology. The *Winnebiko II* was architecturally inflexible, computationally anemic, fragile, difficult to service, limited to a single poor LCD screen, underpowered, and, well, boring by current standards. It was time to do it right!

"Hey, wanna go for a bike ride?" Steve to Maggie during their second date.

Maggie Victor hiking in the Grand Tetons during a roadside stop.

The Other Woman

Ithaca, New York
13,231 miles
August 21, 1987

The road, once begun, never stops. Tonight in Ithaca, over a fine dinner of linguine and wine, coffee and sorbet, we spoke of travel. Hitch-hiking marathons of 17 years ago were revived in candlelight; freight-hopping adventures came to life with all the gritty romance of swaying cars and rusty steel. Bright-eyed and wine-lubed, we recounted nights alone in Nebraska, Jersey cops, long-haul truckers, perverts, road romances, the delicious filth of coal cars across the great prairies, the strange improbable madness of travel… those moments that linger rich in the memory like first loves and last good-byes.

I'm not sure what started my addiction — how I graduated from small doses of bush-league adventure to the hard stuff. I do know that the type of movement doesn't matter very much. The megabike is appropriate and endlessly entertaining, but it's not the heart of my wanderings. The heart is a wild throbbing thing of thermos coffee, road maps, strange eyes, and exploratory kisses — of faded packs and mountains, harbor smells and camp stoves.

No, I have no idea how it all started, but I do know I'm a victim — a happy victim, more willing to pedal over hills on a 95-degree summer afternoon than endure an air conditioned office at $50,000 a year. The road, the Other Woman, is the love of my life… and I'll vow to kiss her sweet asphalt forever if she'll keep me free from the torpor of stability.

She's a tough one, though. This is never an easy relationship. The Other Woman seduces the unwary and has thousands of lovers… yet is jealous of every one. She'll kill you with your own passion or ignore you in some backwater until you scream in frustration. She'll fill you with delicious fantasies then spill your blood without remorse. But still you can never leave her – only withdraw for a while to lick your wounds and sample someone stable, secretly browsing your Rand-McNally in the bathroom like a worn copy of *Penthouse* while dreaming of your next escape. The Other Woman lures you back, time after time, lures you back into her long winding arms like the helplessly lovestruck suitor you are. For once you taste her charms, you are forever spoiled, forever ruined – doomed to fidget through your static spells and gaze misty-eyed at old boots, stir at the sound of pre-dawn freights, pick up hitchhikers in tight-chested jealousy and try not to show your pain.

Yeah… if you don't go running back, you suffer forever.

We're a sort of family, spread across the planet like a scattered clan with a rare genetic disorder, drawn together in common need, recognizing each other in crowds. We are the victims of the Other Woman. We gather around campfires, trade food, grin across the highway with weathered faces crinkled and arms upraised. In hostels, our strange accents tickle each other's ears; we trot out our memories and photos to share insights into what makes the Road the irresistible Siren she is. We can spot each other at a distance, and even sense the stirrings of puppy love that doom the occasional child to a life of wandering — the child who stands on an invisible leash at town-edge, holding his bicycle, biting his lip as we roll past him toward the mysteries of the open road. We wink, knowing

The Other Woman

the moment has been branded onto the surface of that young brain, searing the delicate cortical tissues into a permanent overlay that will subtly alter everything he sees, forever. A future brother...

It's not all men, of course—don't start waving red flags of feminist outrage at those personal pronouns. Women are struck too: just as addicted, just as seriously ruined by the Road for anything even approaching long-term stability. The Other Woman is quite happily bi, luring beauty into her lair, terrifying parents, turning career women into healthy backpack-toting hostelers who push past their road-fear into a life of adventure. They're rare, radiant females, glowing with the flush of urges fulfilled and moving with the free grace of animals... not the stylized grace of fashion.

But as infinite as the Other Woman is, there are certain things she can't do very well – things that leave one fleeing her arms for those of flesh... then returning again and again, running to and fro in confusion like a child caught in a divorce. For years I traveled like that, pedaling from romance to romance against a backdrop of the road. It became a sort of rhythm, a soft succession of new loves, a Russian roulette of pathogens. I would pedal into town and meet her. You know, HER. Eyes would lock. Hands would tremble. She would be drawn into my writing, my bike, the adventure of my life. I would be drawn into her beauty, her warmth, her modular phone jack. Needing a place to stay and sensing the stirrings of passion, I would move in.

By unspoken agreement, the bike would become a piece of abstract sculpture standing in her living room instead of an ominous poised symbol of my transience. The love would grow, fragile, accelerated by circumstance, a whole relationship compressed into days. But then the Other Woman would begin whispering from the dark, and I'd start gathering the Ziplocs, tweaking the bike, scanning my list of contacts. I'd break the news, and try in vain to soften the pain... my chest aching at the tears of sorrow and reproach glistening like jewels on the cheeks of a new friend. Promises... to write, to rendezvous, to remember. And then the last kiss, so terribly different from the first.

Alone, I'd slap on the headphones and crank up the jazz, reset the Cat-Eye and flee back to the Other Woman, that bitch, the rhythm of my pedals salting the open wound of young love shattered again. It got old after a while, the novelty obscured by the pain.

And so we come to the present. I'm rolling around in a menage a trois now: Maggie, the Road, and I. This might be it—a blend of comfort and adventure, flesh and asphalt, love and addiction, freedom and security. The endless changes of travel keep the moss off our toes, yet we suffer not from road-ache, that affliction that renders the lone traveler somehow tragic and driven, a free electron looking for a covalent bond. We've become a molecule, Maggie and I, drifting together from family to family, more a part of the solution than of the precipitate. It's a good life, and I'm even learning to handle the once terrifying stability of a long-term love.

We share road food, conjured from her bicycle trailer by magic. We zip our down bags together to chase the evening chill—our porta-condo a cocoon of healthy smells as the familiar fabric walls billow gently in the breezes of a new place. The rhythms of movement beat like an undercurrent of congas in the night: heart thumping, pedal pumping, file dumping. New towns roll into view, effortlessly, each a haven of new friends and warm beds... each a different view of the same essential home. I write, add bike systems, and expand the family. And it's so easy, this nomadic life, now that the desperation is gone and the tools are familiar.

The Other Woman wasn't expecting this domestication, but she doesn't seem to mind. She still throws us curves, owns our hearts, and leaves us panting... hungry for more.

That's the way she likes it.

The Grand Turing Machine... *Why?*

So what was the objective here? Was this just another step in the layering of progressively niftier technology onto my faithful old recumbent bicycle frame until I'm too old to huff it up a hill? Was there a central dream? Or am I just an inveterate gadget freak, propeller-head, and yuppie hobo all rolled into one, too set in my wandering ways to consider settling down?

These thoughts were much on my mind as I pondered the design of a new machine. The scope of this one was truly frightening: nearly $1.2 million worth of bicycle, including the value of professional time (at 580 pounds loaded, that's $129/ounce). Three and one half years to complete. About 150 corporate equipment sponsors along with a couple dozen occasional assistants, machinists, technicians, and consultants.

What could possibly motivate me to concentrate so many precious resources into a ponderously heavy machine, just to expose not only it but *my own body* to the violent drunken terrors of the highway?

The initial need to escape suburbia for a life of adventure had long since been satisfied. So had the basic tire itch, the lust for a piece of asphalt, the curiosity about hot fleeting romances and the mysteries of small towns. I had already proven to myself that I could "do it," whatever that means. So what was it this time? Inertia? Addiction to movement? A sense of not having finished the job yet? Vague dreams of another book? The continuing appeal of the spotlight and its seductive spinoffs? A lifelong love of whiz-bang technology? It was a little of all that, but much more besides. Consider the motivations...

Let's not forget the REAL objective here. All the motivations mentioned in the text notwithstanding, the fundamental goal is freedom. (Photo taken from Mt. Avery, near Crested Butte CO)

Consider the:

Suite of Motivations

Lifestyle Prototype

Future society will be virtually paperless, energy-efficient, dependent upon wide-bandwidth networking, and generally cognizant of global perspective through routine communication across decreasingly relevant borders. It is not too early to prepare for this: we need the ideas, the tools, and an awareness of the problems that accompany fundamental shifts in the meaning of *information*. The new bike, and the lifestyle that results, was a case study and feasibility test for much of this. I had already seen the effects of my earlier travels, especially back in 1983-84 when laptops, online services, solar panels, and recumbents were all so strange that people were startled into understanding. During the third trip and follow-up speaking tours, I appeared occasionally at schools to help plant the subversive seeds early (reminding students that the obvious choices are not the only ones).

Consulting

Industry is requiring increasing specialization of its workers, due to the overwhelming amount of expert knowledge associated with every technology. This yields positive results but at a severe cost – specialists inevitably lose sight of the big picture. There is thus a growing market for people who travel continuously *among* specialists, cross-fertilizing at every stop. No trade journal or annual conference can accomplish as much as a renegade cadre of curious technoid generalists, on the loose in industry and addicted to surfing the knee of the learning curve. The companies that recognize their own necessarily narrow focus and take steps to keep it in context gain a competitive edge, and my nomadic lifestyle and extensive support technology keep me in touch with a very wide range of pursuits… and thus marketable… a built-in fallback if I ever burn out on living and sharing my technomadic adventures.

Product Potential

The new system addressed a number of basic needs: autonomous power generation, global communication, soft-architecture real-time control, nomadic publishing, security… and more. These needs are by no means unique to me, and casual market research suggests that there could be a wealth of spinoffs with the right strategic partner. Being a business wimp, I've never taken advantage of this, but hey, the thought is there.

Writing & Publishing

People are historically fascinated by life on the edge, the adventures of travelers, and peeks through curtains into other lives. Travel and writing are thus inextricably linked, and by carrying the most sophisticated tools available for biketop publishing, communication, and information-gathering, I minimize the number of excuses for not being productive while raising reader curiosity in the process. For an author, this whole gambit is a gold mine: endless story material, superb tools, and easy marketing based on a recognizable image.

Adventure

This goes without saying. High-tech nomadness is *fun*, and my travel style ensures interesting contacts in strange places. Routine life is impossible on a computerized recumbent with solar panels and a thicket of antennas… and there's a LOT of world to explore out there. Having had a taste of it, how could I spend my life in one place?

Security

I once wrote that "the greatest risk of all is taking no risk." While that may be true, it does not mean that I relish the idea of being robbed, run over, or left to die of thirst with a broken axle in the desert. *BEHEMOTH* was designed with enough different kinds of communication gear to virtually ensure that I can get a message out if necessary. Usually help is only a pushbutton away, via ham repeater or cellular phone – in more remote areas, a few moments' preparation puts me on the HF bands or into a communication satellite.

Global Community

This may sound odd at first, given the classic "loneliness of the long-distance traveler." But nomadness is the most social lifestyle imaginable for two reasons: global networking and the timeless energy of beginnings. The traditional concept of stability, normally restricted to neighbors, associates, and the familiar things that define "home," is now distributed around the world and constantly refreshed by encounters on the road. Home is everywhere, and I am constantly amazed by the intelligence and imagination lurking in the most unlikely places. (The mainstream high-tech world is provincial and technocentric… seldom recognizing that wizardry can thrive even in backwaters not steeped in the vapors of silicon. From unexpected quarters come new ideas.)

Technical Challenge

Finally, one of the most deeply alluring parts of this whole affair is the project itself. The whole system borders on madness, so diverse and intertwined has it become. The engineering aspects of this, ranging from sophisticated CAD tools to fancy new adhesives, represent a seductive and multifaceted learning curve coupled with the pure joy of creating something exciting. And one of the best parts is that the media visibility keeps attracting new sponsors, allowing me to select the very best technology that industry has to offer without being stopped in my tracks by something so mundane as cost. How could a dedicated hacker/tinkerer ever abandon such a project? The *Microship*, which is nearing launch at this writing, carries this whole notion to even wilder extremes—it's a technological addiction, spiced with high adventure.

So how do we turn all this into a design specification?

Adventures in South Ecotopia

November 28, 1986

Do you ever read my stories and wonder what it REALLY feels like to be out here, exposed to the world, unsure from one day to the next where I'll sleep, who I'll meet, what pleasures and pains will strike with the whim of chance? Do you ever try to see past the rhapsody, the humor and philosophy—looking for clues in the rhythm of my words, sensing exhaustion in torpid prose or the giddiness of new friendship in silly sentences of puns and alliteration?

Narrow-bandwidth communication like this is frustrating. I'm living an adventure of intense visceral sensation, and the only way I can share it with you is through words—and maybe a stack of photos if I ever camp in your livingroom and swap tales over pizza. Not enough. Today I wanted to share more: I wanted you to BE there.

It wasn't a normal day, this 18-mile explosion of violence and insanity. It was a day of curses lost in the spray of trucks, of stinging eyes and cold sweat. It was a test of hardware, a test of nerves, a challenge to muscle and mind alike. Today was one of those days that will live on as a caricature of the entire journey—a day that will instantly spring to mind whenever anyone mentions riding in the rain… or redwood trees… or the sheer looniness of challenging truck-infested mountain roads on a bicycle in a heavy storm.

Imagine sweat, lots of sweat, steaming inside layers of polypropylene and Gore-tex. Its pressure builds, hot and stifling, as you strain in a headwind up a mountain road. You think to disrobe, but the icy trickles of rain leaking through zippers and seams warn otherwise—better to be hot and wet than cold and wet. Your shoes begin to squish, and you make a fist every few minutes to squeeze water from expensive "waterproof" neoprene gloves.

Soon you accept the discomfort and pay more attention to the other problems: packs soaking through, computers and humidity, trucks blasting by in an opaque spray. Those can be challenging as you waver unsteadily up the grade at 3 mph, fighting crosswinds. Sometimes they catch you broadside in a soaking explosion of white water and roar off into the mist, trailing diesel fumes and the smells of chopped fir, leaving you struggling for control as a motor home passes too closely and a knot of vegetation forces a swerve into traffic. Ah, recreational cycling. The water is everywhere—inside you and around you. You need to vent the morning's coffee, swilled so long ago in a fluorescent-lit '50s cafe, but the grade is too steep for parking… so you press on into the rain, splashing in brown runoff like a spawning chinook, pedaling numbly and dumbly and trying not to think about the place you could have stayed a few miles back. Giant trees pass slowly, shrouded in mist; the sounds are a muted cacophony of patter and splash, drip and roar, bicycle chain and your own wheezing breath. Higher you go.

And then the summit, understated, no sign but a warning to trucks, no place to pull off and congratulate yourself. Without fanfare you coast the level part, breathing easily, relaxing slightly—then your speed picks up and the curves fly by and the bumps are terrifying… the brakes are wet and your hands grow numb… raindrops sting your face and you squint into the gray, peer into the murk, scan the blurred submerged pavement for signs of potholes and glass and ruts and bumps and—*HEY! GIMME SOME SPACE, JERK!* —anything else that could drop you in a blink and spread you like a high-tech road kill across two lanes of uncaring violent glorious redwood highway.

This is the kind of cycling that makes the first motel look like a sort of paradise. You hand over a dripping Visa card then drag your bike inside, spreading wet fabrics over every door, chair, and light fixture—steaming up the room while lying numb and smiling in a real bed. What a life…

And I wouldn't trade it for all the BMWs in suburbia.

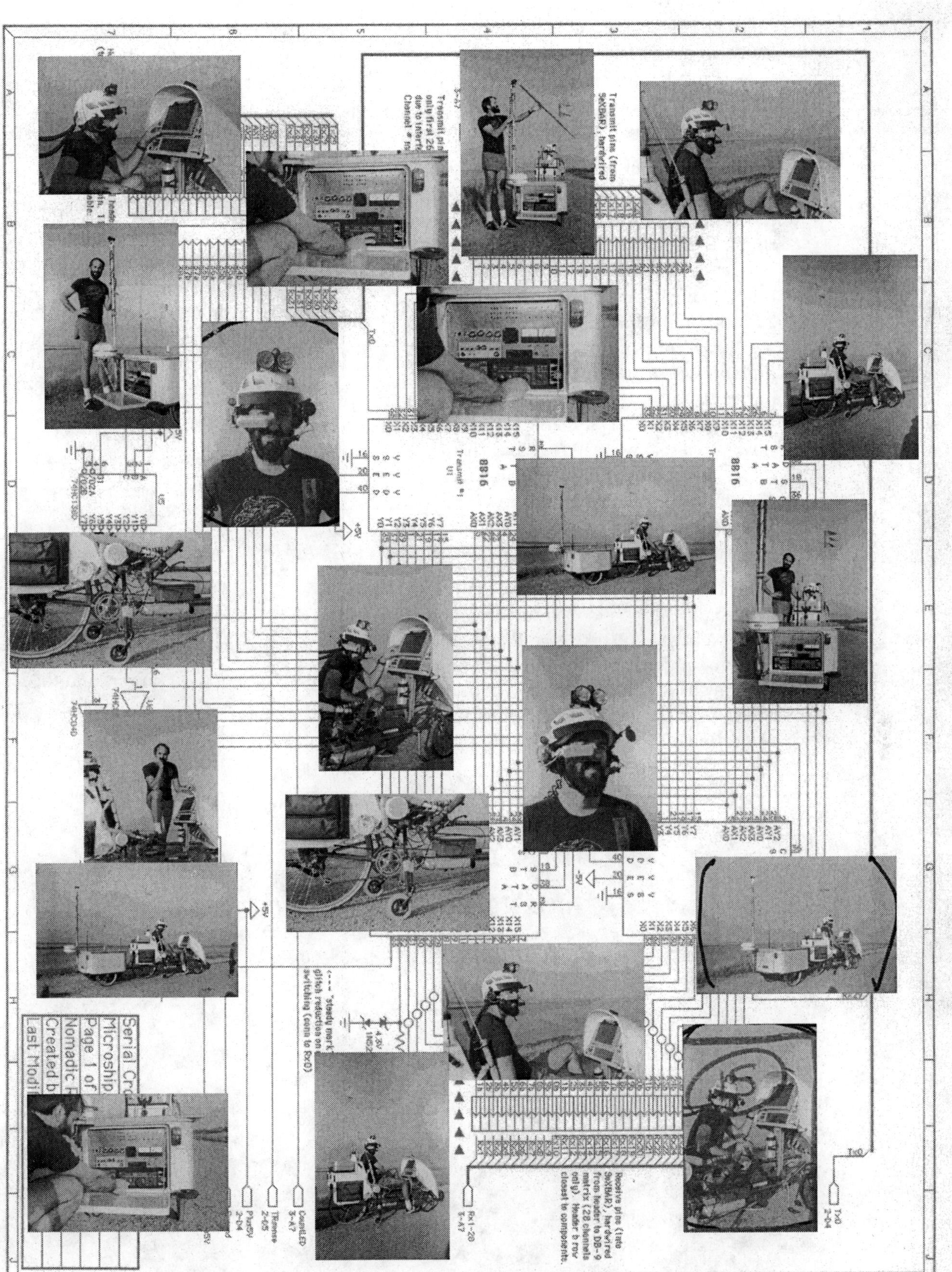

Chapter Two
BEHEMOTH Technology

BEHEMOTH THE MEGACYCLE IS BORN

Okay. Given all those motives, a general set of objectives became obvious for the bicycle that became known as *BEHEMOTH* (Big Electronic Human-Energized Machine… Only Too Heavy).

While I cannot claim that all of those goals were adequately met before I declared the project frozen in 1991 and moved on to shipbuilding, the 3.5 year development project yielded a system that embodied much of the spirit of the above. Let's take a quick tour of *BEHEMOTH* from the perspective of my time living aboard. (This little disclaimer necessitated by the staggering technological changes since then – if I sound like I'm rhapsodizing about ancient slow processors with 40 megabyte hard drives, it's because I'm pasting in a technical description written when this machine was a state of the art tour de force of Silicon Valley gizmology!)

First, I should give you a quick physical description – take a look at the image on the next page. The bike is an 8 foot recumbent, meaning that I sit in a relaxed position with my hands on control grips at my sides and my feet latched into a crankset out in front. Behind the bike is a four-foot yellow trailer with solar lid, flip-down communications bay access door, and numerous antennas. In front of me is a large control console contained within a smooth white Lexan fairing, presenting a panel with three large LCDs and numerous other instruments. Behind me a large white "RUMP" provides additional equipment space, a helmet cooling system, more antennas, and a docking bay for an aluminum manpack with its own small solar panel. Atop my head is a decidedly bizarre helmet with heads-up display, motion sensors for cursor control, lights, the fluid heat exchanger, and an audio system. The whole system, fully loaded for touring, weighs about 580 pounds… plus me… and thus has 105 speeds, deployable "landing gear" for mountain climbing, and hydraulic brakes to help me survive the descent.

BEHEMOTH is deeply integrated, but can be considered in eight major chunks: power systems, computers and human interface, communication and security, equipment packaging, the R&D lab, the bike itself, camping/support equipment, and the base-office infrastructure.

But first, the basic design goals…

> *"There's glory in using inappropriate tools. You can tell you're pushing a new frontier when all available tools are inappropriate."*
>
> *– David Berkstresser*

SO WHAT ARE WE TRYING TO ACCOMPLISH HERE?

The machine will be an efficient human-powered vehicle optimized for comfortable long-distance travel with minimum external support facilities.

- It must carry state-of-the art mechanical, safety, medical, life support, and camping equipment.
- On-board systems must provide maximum possible autonomy in electrical power generation, computation capability, file storage, navigation, and communications.
- All components must be easily maintainable with on-board CAD tools and R&D equipment, and all system documentation must be machine-readable or microfiched.
- Operation while riding and while fixed must be supported equally, and both modes should maximize communication bandwidth between man and machine.
- On-board computers must support major application software standards, and the embedded control systems must be adaptive and power-efficient.
- There should be a minimum number of hardwired switches, with all control and management functions depending instead on an extensible graphical user interface (except for basic safety equipment).
- Future hardware additions or upgrades must involve minimal hardware changes.
- No radio operations should require antennas with external supports.
- The security system must be robust enough to allow me to turn my back comfortably for more than a few minutes.
- And it must be beautiful, fun, and in keeping with current standards of engineering elegance!

BEHEMOTH near Sun Microsystems, Mountain View CA 1991. (Photo by Maggie Victor)

Power Systems

One of the primary design specifications noted above refers to "maximum possible autonomy." Most commercial electronic systems assume that power is freely available, and thus exhibit rather sloppy management of loads (comparable to automotive and building design before people started thinking about fuel cost).

I don't have that luxury. AC outlets are often days apart while touring, and there's no convenient gasoline engine to keep a battery charged. Power, like my own human energy, is thus a scarce resource, and a fair percentage of *BEHEMOTH's* control system is devoted to its management. Curiously, this is one of the reasons there are so many computers: a good multitasking system could handle everything I need to do, but it would have to be active – and drawing lots of power – all the time, even when only monitoring a security sensor bit or two.

There are three power sources, one on the bike itself and two external. The primary source is a 72 watt array of Solarex photovoltaic panels, which forms the lid of the trailer. A small ten-watt panel graces the lid of my manpack, theoretically allowing me to unplug from the bike and still maintain reduced computing and communication capabilities *(although in practice, the SPARCstation ended up in that case, far exceeding the capabilities of a ten-watt solar module).*

The other two power sources are the external AC line (any international flavor) and an external source of automotive-scale charge current.

A pair of 15 amp-hour Sonnenschein sealed lead-acid batteries are in the basement of the trailer, and another is in the RUMP (Rear Unit of Many Purposes). An automatic change-over circuit allows trailer disconnection without glitching sensitive loads.

Power management is in two layers. If no computers are alive, a charge regulator keeps the batteries from overcharging. But Hall-effect current sensors at all key nodes allow one of the microcontrollers to keep track of current flow, redirecting excess power to optional loads or simply disconnecting it when the batteries are full.

The main bike power bus is 12 volts, but electronic loads call for lots of other voltages (mostly +5). To produce these as efficiently as possible, Power Trends integrated DC-DC converters associated with each load are switched in as needed, yielding 85-90% conversion efficiency and a minimum of unneeded overhead. These are FET-switched by the trio of control processors as their associated subsystems are invoked by higher levels.

On top of all this, a Mac-based graphic front end can display a dynamic pictorial representation of the whole power system, showing battery levels, all charge and load currents, and projected time until either full charge or total depletion (based upon sliding averages and historical trends).

Seventy-two watt Solarex array on trailer roof, tilted for optimal exposure. BEHEMOTH is in early development here. (Santa Cruz CA 1989)

COMPUTER AND CONTROL SYSTEMS

This is *BEHEMOTH's* heart, or perhaps I should say hearts. When everything is active, there are a number of microprocessors working in harmony – some as highly localized embedded controllers, some as bike-management systems, and some as high-level applications platforms. I will not attempt a complete inventory here, but an architectural overview is in order.

The primary user interface to the bike is implemented in HyperCard on a repackaged Macintosh Portable, whose hinged screen is at the top center of the control console. Mouse pointing can be done with head movement (via ultrasonic sensors and a dedicated controller) or via a conventional mouse plugged in when the bike is stopped. Likewise, data entry is via the handlebar chord keyboard that serves all machines, or a plug-in QWERTY keyboard that lives with the mouse in a pack behind the seat. When underway, synthesized keyboard and mouse events are passed to the Macintosh via the Apple Desktop Bus, eliminating any need to hack the Mac logic.

HyperCard and the Finder run in RAM disk, minimizing the need to spin up its 40 megabyte hard drive *(remember those?)* while mobile. All bike control is ideally represented through a variety of "views," each bearing graphic elements that are linked through external commands to the bike's trio of control systems, layering the smooth Macintosh mouse-driven user interface atop the FORTH command lines. These control views are divided along application boundaries, such as security, power, network configuration, maintenance, data collection, ham radio, packet datacomm, and so on. The point of all this is complete flexibility in interface design – the precise opposite of the *Winnebiko II*, whose hard-wired console was virtually impossible to reconfigure. *(As with many other parts of BEHEMOTH, we ran out of time before this was fully finished, but the concept was proven well enough that it is now being implemented on a much larger scale in the Microship.)*

Other Mac applications include biketop publishing, electronic mail, word processing, communications, MIDI system control, and a slide show of graphics and scanned "electronic sponsor decals" for use when *BEHEMOTH* is on public display.

Flipping up the Mac screen reveals a Sharp VGA LCD driven by an Ampro Little Board 286 system, essentially a 16 MHz AT-class DOS PC packaged for industrial control and drawing about 8 watts (including math coprocessor, 4 megabytes of RAM disk, 3.5 inch floppy, and a second shock-mounted Conner 40 meg hard drive). The backlit screen allows any robust DOS graphics application to execute perfectly well on the bike. This translates into easy use of OrCAD, Autocad, programmable gate array software, computer-generated maps, and much more... though here, the reality of a backlit passive-matrix screen under high ambient light conditions has rendered this PC almost useless on the road.

A second and more useful Ampro DOS system, located behind the seat, runs a number of applications including Reference File, a pop-up database manager that allows me instant access to about 5,000 contacts around the world. Since this is in frequent use on the road, it lives in the system that drives the heads-up display on the helmet – Reflection Technology's Private Eye. This 720x280 screen presents the illusion of a floating display just below my center of vision, no matter where I happen to be looking (even in rain, darkness, or direct sunlight).

The console carries a third DOS system, based on the innards of a Toshiba T1000. It may seem redundant, but remember the power-management problem – often, all I really need is a simple large-print text editor or communication program. This can run a utility called PRD+ that is piped into the text path from the handlebar keyboard, allowing me to type in a sort of shorthand to more than double my effective data rate. It also handles a Covox speech-recognition

Computer and Control Systems

system, enabling voice command of basic bike functions via helmet or remote radio. *(But not continuous text entry, not in 1991 with an 8088 CPU!)*

I also haul yet another machine in the form of a laptop (a Macintosh PowerBook 170) in the trailer. This allows routine text work and software development when off-bike with the manpack, and also supports full remote control of the on-board systems from up to 2 miles away via a business-band packet data link (described below).

Finally, just before departure we added a Sun Microsystems SPARCStation to *BEHEMOTH's* applications layer. I have always wanted to ride a unixycle, and besides… that kind of horsepower is an excellent environment for map generation and computation-intensive CAD. It also makes the bike a roving TCP/IP node, tied directly into the Internet via a Telebit CellBlazer modem and Morning Star PPP software.

BEHEMOTH's control console, featuring Macintosh screen. (Photo by Bill Brown, WB8ELK, Peterborough NH 1992)

The control environment includes multiple processors, with the stars of the show being a trio of New Micros 68HC11 systems. These low-power machines wake up speaking FORTH from ROM, making them exquisitely hackable without external development systems or the traditional edit-compile-debug cycle, and they are I/O-rich and slow-clockable to minimize current drain when the computation load is light.

The main machine here is the BCP, or Bicycle Control Processor. It has, in addition to its console port to the Mac, a pair of serial ports on dedicated lines to the RCP and TCP – similar systems in RUMP and trailer. Even though a single 68HC11 could handle all the control tasks on the bike, doing it this way significantly reduces cabling and complexity.

These controllers handle everything: load switching, data collection, security, handlebar keyboard decoding, power management, and network configuration. One of the most significant tasks is managing the crossbar switch matrices through which pass all audio and serial communication in the machine. Within each major site (console, RUMP, and trailer), a Mitel 8816-based matrix is wired to all local devices. These are tied together by 12 shielded "long-lines" (8 audio and 4 serial), which are dynamically allocated to communication events as they arise. An incoming cellular phone call while mobile, for example, could result in the phone's speaker and mic audio being assigned to buses, which are then switched to the helmet's boom mic and left earphone (while the right ear may remain connected to the Audapter speech synthesizer or other audio source).

Again, the whole idea here is flexibility. I could not possibly anticipate all future enhancements at design time, so a system architecture that treats every device as an addressable node with a minimum of dedicated interface circuitry is most likely to remain useful for the life of the machine.

The Brain Interface Unit (BIU)

First published in Nomadness, Issue #9 Fall 1990

On the road, *BEHEMOTH's* bio-controller is always embedded in its modified Bell Tourlite shock-isolation shell, interfaced through visual, aural, and kinesthetic channels to on-board silicon-based systems. I'd like to give you a brief overview of BIU functionality...

Naturally, every effort has been made to maximize communication bandwidth with the neuron-based system inside the flesh-shrouded head assembly. A 720 x 280 pixel display presents an apparent graphic overlay upon the system's binocular view of the world, spectrally peaked at 720 nanometers to minimize any ambiguity with reality and adjustable in apparent focus to minimize attention-switching stress. A second visual sub-window is provided by a circular optical reflector mounted on the solar attenuation shield, giving the controller a steerable view of conditions aft. Optional visual attenuation filters can be installed under conditions of high solar flux, softening specular reflections while diverting airflow-borne particulates from the moist and delicate components of the image-acquisition optics.

Both of the rider's aural channels are coupled to transducers that allow reception of synthesized human language, long-range bidirectional RF communication with others of the same species, alert messages, or any of a number of laser-extracted cross-interleaved Reed-Solomon digitized stereophonic wavefronts selected for relaxation, stimulation, motivation, or subjective time-compression purposes. Note that these transducers are of limited bandwidth, but can be augmented by miniature units inserted directly into the auditory canals, or alternatively by high-power acoustical drivers located behind the entire brain packaging system. Wetware initiated lexical utterances are converted into analog data by a boom-mounted input transducer, and are coupled through the audio network to speech recognition, recording, or communication subsystems as required.

Due to the human visual system's insensitivity to infrared and other useful wavelengths, the BIU incorporates powerful parabolically-focused light transmitters, with two different degrees of luminance and collimation to accommodate varying conditions. This has been proven more effective than constraining the beam's axis to that of the substrate vehicle, since the bio-system is capable of quickly adjusting the azimuth and elevation of the entire head assembly to center the region of interest in its visual frame of reference (not necessarily co-axial with the current physical trajectory). Whenever the system is traversing a region of the planet that is devoid of insolation, reflections of these beams from landscape features and signage allow real-time decision-based navigation at normal velocities. We considered reducing power requirements by overlaying an image-intensification system upon the visual field, but this is not an effective solution... it is highly beneficial for autonomous wetware systems piloting petroleum-based land vehicles to recognize the presence of *BEHEMOTH* and take appropriate evasive action.

The ability of the bio-system to track objects of interest through precise 3-axis positioning of the head assembly enables

The Brain Interface Unit

an additional level of interface with the on-board computer network. Three 40 kHz ultrasonic receivers positioned on the BIU's crown and temples receive a reference beam transmitted from the console. Pitch and yaw angles are derived from raw phase and doppler information, and are used by a dedicated processor to determine precise head pointing angle. These data, in quadrature form, are converted to conventional ADB events and passed to the Mac, yielding an apparent link between the rider's nares and the on-screen cursor. All mouse pointing is done with gross angular BIU movement; clicking and dragging are accomplished via handlebar contact closures antipodal to the data-entry keys associated with the user's phalanges.

The BIU is designed to cushion the relatively delicate host organism upon occurrence of rapid deceleration associated with impact. Should the human system separate from *BEHEMOTH* and become launched upon a divergent ballistic trajectory, the twin coiled cables carrying all interface lines will achieve full extension, actuating lanyard-release connectors. This is designed to prevent abrupt cervical misalignment (or separation) in high-velocity crisis situations.

Biophysical temperature rise resulting from the accumulated losses of propulsion workload (exacerbated by the low thermal-conductivity of the shock-isolating foam shell, especially under conditions of elevated-ambient) can be controlled through a fluid heat exchanger closely coupled to the scalp and thermally cycled with an insulated reservoir of aquatic phase-change nodules via a manually-activated peristaltic pump. Bio-system hydration is managed via a small bite valve that can be inserted into the rider's speech output device (which doubles as the input port of the alimentary tract and auxiliary oxygen-uptake unit).

Since the osseocarnisanguineoviscericartilaginonervomedullary system is essentially dependent upon fluid-evaporative cooling involving a large percentage of its epidermal surface (despite significant augmentation by the cranial heat-exchanger), there is potential for heavy accumulation of concentrated saline exudate within the crushable BIU interface layer. This is reduced through an absorptive accumulator that can be manually cleared, as well as a circumferential fluid-removal channel that carries waste coolant back to the occipital region and thence into an overflow tube terminating at the first thoracic vertebra. These measures insure a minimum of irritation to the delicate ocular membranes under heavy load (the wetware information system, though only dissipating about ten-watts, is unfortunately dependent upon the same metabolic processes that support the bio-engine's fuel, waste, and heat management facilities).

In short, the BIU is the key interface link between *BEHEMOTH* and all aspects of its biological host organism. It provides crash safety, cooling, hydration, sweat removal, visual graphics display, luminance attenuation, communication and entertainment audio, a voice control channel, a view of the road behind, a steerable light source, and a mechanism for hands-free mouse control.

And I never ride anywhere without it.

Communication and Security

Computers are wonderful, but if the bike were an island of processing power unconnected to the world its utility would eventually seem questionable. Publishing, network communications, security, ham radio – all depend on links between *BEHEMOTH* and people (or machines) *out there*.

Strangely, one of the features of the *Winnebiko II* that most appealed to people was the cellular phone. Perhaps this is just because it happened to be one of the few recognizable components, but I prefer to think that it's because there is a fundamental human fascination with staying in touch while moving around. (The Oki cellular phone is no longer recognizable… it was depackaged and integrated into the network like everything else. Who needs all that excess plastic?)

The cellular phone has become a ubiquitous part of business communications, and a number of interesting products have emerged to take advantage of it. I carry both a Microcom Microporte (MNP-10) and the CellBlazer modem mentioned above, allowing communication with traditional wire-line services while mobile or camped away from RJ-11 modular jacks. And should I need a modular jack for something, I can plug into the cellphone via a product called the Celjack. This allows the use of fax, a credit-card verifier, and an answering machine no matter where I happen to be. Naturally, I also carry standard modems for use when twisted-pair phone lines are available (as well as acoustic cups for pay phones and a pair of alligator clips for motels that are still in the dark ages).

Of course, cellular phones require users to be within range of a transmitter site, and they also cost quite a bit to operate. The first problem has been partly alleviated with the addition of a six-element yagi beam antenna from Larsen, but a much more dramatic long-term solution is the OmniTRACS satellite terminal from Qualcomm. This one-watt spread-spectrum 14 GHz transceiver gives me a slow but reliable 24-hour Internet connection anywhere in North America via the GTE GSTAR satellite and Qualcomm's Network Management Center in San Diego. A fully automated network gateway links mail transmitted via this path into the Internet, and presents my mail in Eudora just as if I were using a modem.

In addition to all that, the bike is well equipped with amateur radio equipment, both voice and digital. Ham radio has the advantage of coming with a fairly cohesive community of users, simplifying the problem of finding help or places to stay in strange towns. As an energetic hobbyist culture, it's hard to beat, and a flood of new developments in high-speed packet networking, satellites, television, and other modes are keeping it interesting year after year. *(Looking back, I cringe at how much more robust BEHEMOTH's communications would be if I were building a new version today… things I fantasized about in my less-rational moments are now off-the-shelf gadgets!)*

Rear of BEHEMOTH's WASU (Wheeled Auxiliary Storage Unit), showing Qualcomm OmniTRACS satellite antenna, ham station, and LED taillights. Qualcomm terminal shown above left. (Photo at a campground in Wisconsin 1991)

Communication and Security

At the "appliance" level, a miniature Icom handheld 2 meter transceiver is repackaged and built into the console, allowing me to chat easily with traveling companions and local hams as I travel (there's a push-to-talk switch on the left handgrip, and the unit is ported into the bike's audio network). I also carry a small handheld when I'm off-bike for pretty much the same reasons. It gives a traveler in a strange town a nice sense of security to key up a local repeater, identify (N4RVE), and find immediate welcome from a local community.

Traditional HF ham radio is supported on *BEHEMOTH* as well, via an Icom 725 mounted in the trailer. This rig has a computer-control port, making it easy to operate mobile using a compact folding dipole fashioned from Outbacker mobile whip antennas. When camped, I sometimes unroll a wire dipole from a pack and operate under more normal conditions, though the on-board dipole has yielded surprising results in the form of relaxed international ragchews and even a contest or two. Whatever the mode, the HF station provides not only the well-known fun of ham radio but also the virtual certainty of finding someone, somewhere, in the event of an emergency.

There's more. Yaesu 290 and 790 multi-mode rigs for VHF and UHF are in the trailer as well, and these not only give me a high power option for long-range repeater use, but also handle packet radio. This digital mode links me to a global wireless electronic mail network entirely owned and operated by hams... and new Microsats add the remarkable capability of making an email connection twice a day from anywhere in the world (given suitable antennas and a PSK modulator). These tools are growing fast, and can even provide relatively seamless TCP/IP links with Internet and other services... putting me in direct (albeit leisurely) contact with millions of computer users worldwide from a campsite in the middle of nowhere, whether via packet or the various HF modes.

The only caveat with ham radio use is express prohibition of business traffic, and I have one application that is clearly in that category: laptop-to-bike data link for manuscripts, control, and correspondence. For this, I use a pair of Motorola business-band UHF data radios with dedicated TNCs (terminal node controllers) from PacComm. These offer an extremely useful feature not normally found in such components—a parallel I/O port. Because of this, I can leave the bike's packet station active when everything else is shut down, then sign on via laptop from a distance and transmit commands that cause the control system to be powered up and patched through the packet link to my remote "console." At this point, I can do anything that is possible from the bike seat: spin up hard disks and transfer files, connect to online services via cellular modem, check batteries, set security functions, make rude noises, talk to passers-by with

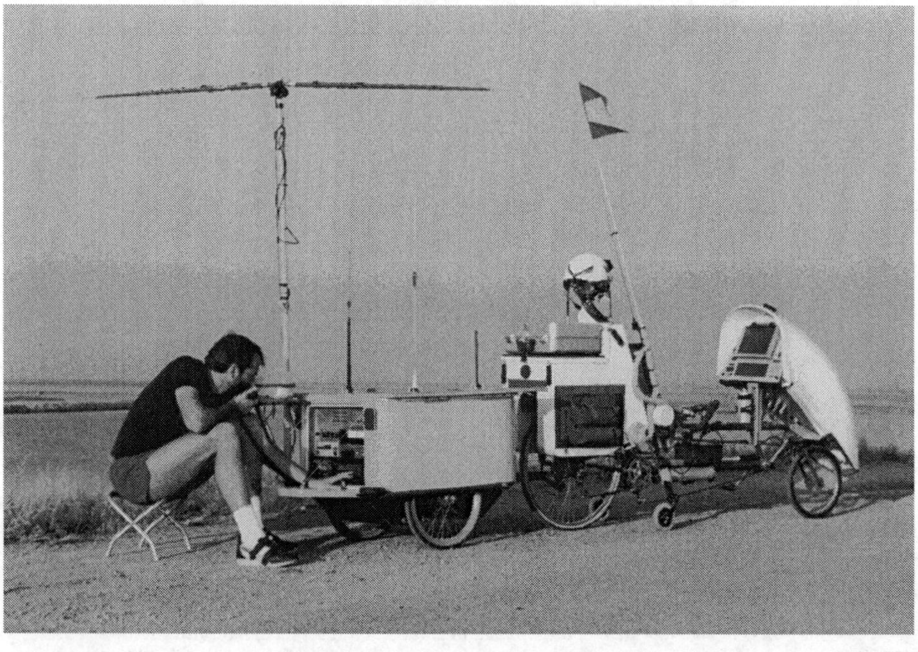

N4RVE working the world from BEHEMOTH.

the synthesizer while monitoring their responses through a wireless intercom, and so on.

The last ham radio component is a small color TV transceiver from AEA, which can tie into my video equipment to allow sharing the sights with a growing community of amateur TV stations. There's just no shortage of experimental gizmology to keep me amused out there...

Security also falls into the communication category, largely because the best lock is the human eye and the basic objective of the security system is to keep me in touch with *BEHEMOTH* even if I'm out of direct visual contact.

There are six levels of security sensors, and a variety of automatic responses to intrusion that depend upon the environment, violation level, and my current paranoia index. When parked in front of a restaurant, for example, human proximity is not an alarm condition—but it certainly would be if the bike were locked in a supposedly secure garage overnight.

The "long-range sensor" is based on an Alpha microwave doppler motion detector. A special conical antenna reflector allows this normally directional device to see about ten feet in all directions from its vantage point atop the RUMP. Level two is physical motion, using the well-proven UNGO Box sensors and associated electronics. If this is triggered, someone is touching the bike (or it's very windy). Level three is the opening of any access panel or the unplugging of any major harness cable, level four is the toggling of any switch on the front panel, and level five is wheel movement. Level six, the most serious, is a change in satellite nav (GPS) coordinates without entry of the proper access code, indicating that someone is in the act of stealing the bike (or that I had one too many beers with my pizza and forgot the password—although with a binary handlebar keyboard it doesn't have to be much of a password...).

The default response to any condition above the current alert threshold is for *BEHEMOTH* to "beep" me via my two meter handheld transceiver. I can then issue a touch-tone command to cause the system to transmit a more detailed spoken alarm with the speech synthesizer.

Microwave motion sensor cone atop ten-GHz feedhorn.

The graphic security management screen allows any number of other responses, again depending on conditions. In some cultures, having the bike speak a loud warning is extremely effective—it denotes ominous machine intelligence in something that is already quite alien in demeanor. "Do not touch, or you will be vaporized by a laser beam!" In other cultures (Silicon Valley), this is correctly regarded as amusing nonsense. Alternative responses thus include a traditional auto security siren or anything else that seems psychologically appropriate. Under more serious alert conditions, the bike can dial 911 and deliver an emergency message ("Hello police, I am a bicycle; I am being stolen; my current coordinates are..."), while beaconing its location along with a call for help on ham packet frequencies and via the Qualcomm satellite link.

Of course, the best part of the security system is peace of mind—my own nervousness, from a classic risk management perspective, has always been much more serious than the actual threat of theft. In practice, there's never been a serious problem.

Geography Falls Apart Online

excerpt from Computing Across America

A life online is a life outside the strictures of physics – at least the conventional physics that make a country seem big. The network is a village shaped like a map of the world, a place populated by spectres and ghosts, travelers in time and space. The physical reality of nodes and satellites is a forgotten substrate, a thing no more in our consciousness than the earth's mantle.

I live there as much as I live here. Since "here" keeps changing, "there" has become home.

Geography falls apart online. It's an odd contrast: I am living two dramatically different lives. One is visceral, sweaty, attuned to every hill and headwind. The other is ethereal, intellectual, an electronic interlocking of imagination and communication. I am at once a being of cloud and soil, satellite and bicycle; I can drop a thought from the sky onto Baton Rouge while patting a white samoyed named Fluffer in a snowed-in Telluride house trailer.

No problem.

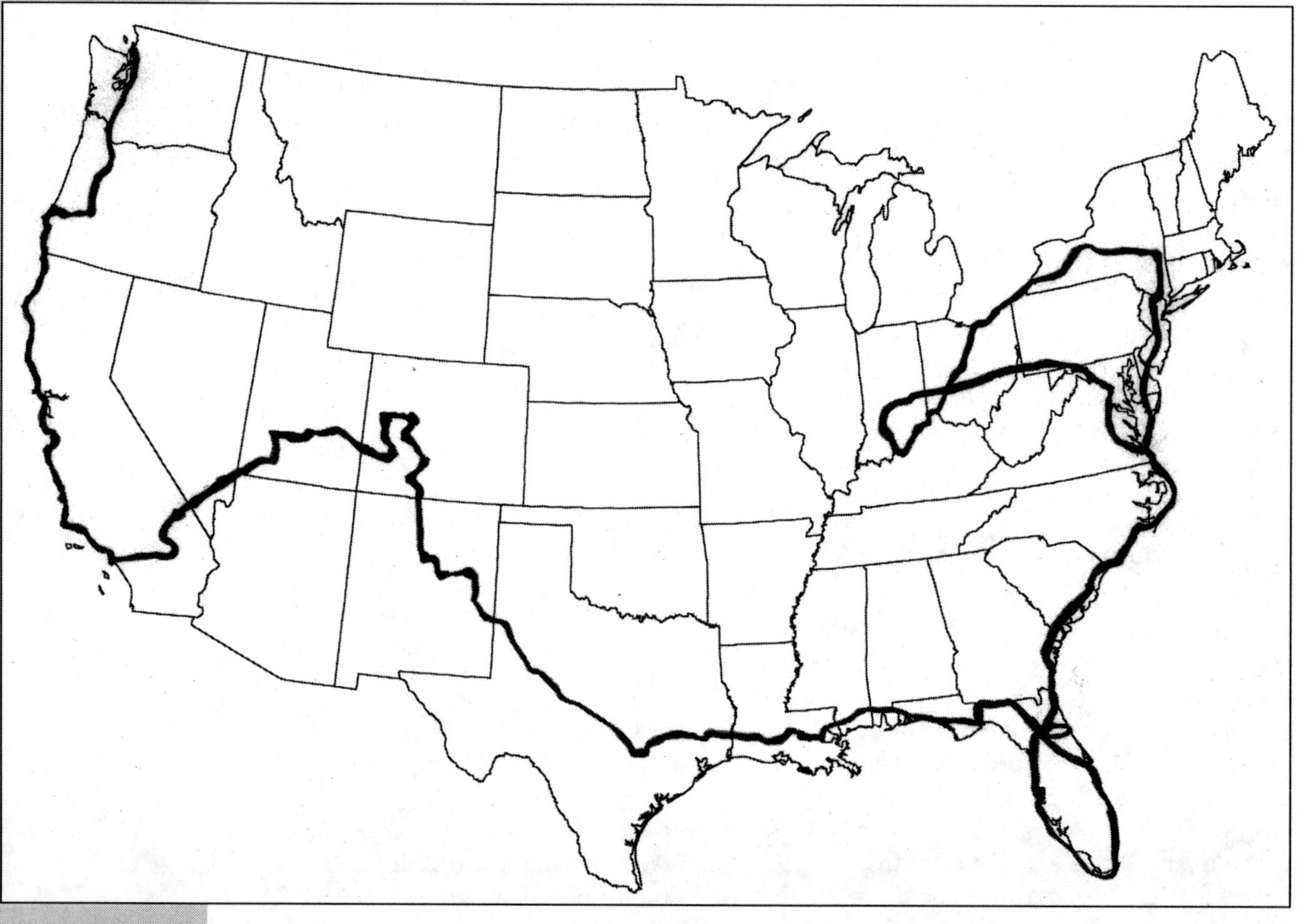

Adventures aboard Winnebiko and BEHEMOTH covered 17,000 miles around the US.

Opening the console – from top to botttom.

1) After removing the fairing, fold down the side panels—Head-mouse controller and Covox speech recognition interface visible.

2) Open forward bay—Multimode packet and Audapter text-to-speech boards visible, as well as Mac battery.

3) Release Southco quick-disconnects holding "main-triangle" in place—Perforated units on left carry BCP and nexus board.

4) Begin unfolding, allowing front panel to hinge down—Macintosh Portable board visible by left hand.

5) Complete opening with all boards accessible—Ampro 286 and hard-disk platform visible.

6) Nexus board (R-N Quick-Connect) folds down to expose New Micros 68HCII FORTH bicycle control processor.

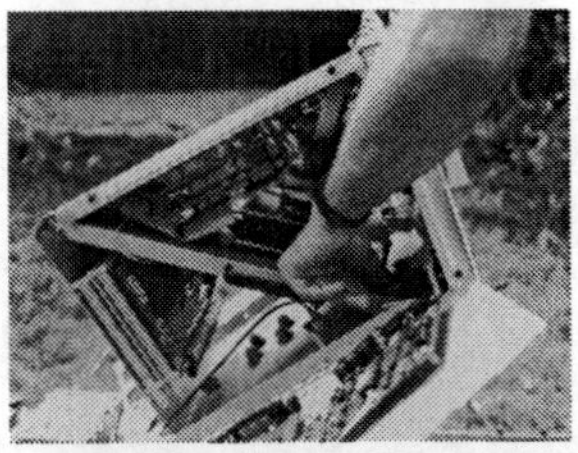

Many thanks to...

- *David Berkstresser, for magical CAD work.*
- *Walt Wilson, for sheet metal shearing and advice.*
- *John Sisler, for help with console sheet metal assembly.*
- *Ken Boetzer, for machining facilities and TIG welding.*
- *Jim Gibson, for the board hinging system.*
- *Ron Covell, for fairing mounts and the hinged Mac screen.*

Equipment Packaging

High-tech equipment is exciting indeed, but there are all sorts of problems associated with installing a wide range of delicate components onto a mobile platform that has to withstand years of heavy shock, rain, temperature extremes, and human abuse—yet still be easily serviceable. Fortunately, circuit fabrication techniques have evolved to the point where the silicon innards of most electronic devices are extremely rugged... but that does not necessarily include the I/O devices, disks, frequency-determining components, and so on.

There are three major equipment bays in *BEHEMOTH*—the console, which unfolds completely for service; the RUMP, with an access cover to a shock-mounted platform; and the trailer, which contains all radio communication gear as well as a secondary operating position. The latter two are constructed of CSPC (cellulose core, silicon matrix, polyester-filled composite—otherwise known as fiberglass over hot-glued corrugated cardboard). This material is light, waterproof, cheap, and easy to build in any arbitrary shape.

The console is where most of the hard-core action is, and it is a complex articulated structure shock-mounted to the frame, protected by both a custom fiberglass fairing and secondary waterproofing layers. Releasing a succession of Southco twist fasteners turns the entire affair into a workbench, with easy access to every subsystem and the back of the control panel... along with such features as a tray for loose parts and a place to plug in the logic probe. Boards that require frequent work are themselves hinged, such as the "nexus" board that carries all random logic associated with the BCP.

To a somewhat lesser degree, the RUMP and trailer offer similar access to internal equipment. But as we move aft, there are more and more RF and power devices, necessitating a different class of design in order to keep noise to a minimum. Suffice it to say that every component is either a custom assembly or a repackaged commercial product, requiring some special attention to keep it working and fixable during the unknown miles ahead.

R&D LAB

Speaking of serviceability, what, exactly, do I service this with? The standard suite of bicycle tools is on board, of course, but that hardly takes care of EPROM programming, glitch chasing, interface hacking, logic probing, analog calibrating, and embedded system debugging.

Nestled among the packs of clothing and camping gear in the trailer is the R&D Pack, a collection of tools and parts that handles most field-development and maintenance tasks. This includes a Createc 10 MHz handheld dual-trace oscilloscope with LCD screen and flexible user interface. Naturally, I also have a small logic probe and digital multimeter—as well as a good selection of precision hand tools.

For wiring, there is an Ultratorch butane soldering iron (which also does heat shrink) and an insulation-displacement tool for the Robinson-Nugent prototyping panels. A healthy inventory of chips, discretes, packaging hardware, special spares, and hookup wire rounds out the R&D Pack.

A number of integrated diagnostic tools are required for a system as complex as this. An Acculex digital panel meter with thumbwheel switch is mounted on the console to allow quick hardware checks of all major system voltages, and a 6x8 array of LEDs on the front panel is driven by a dedicated PIC controller (designed by Dan Kottke) that allows display of 48 individual bit conditions including power-control lines. This is normally off to conserve power, of course, but when things aren't working it provides an instant window into *BEHEMOTH's* internal organs.

And then there are the development environments—FORTH for the controllers and HyperTalk for the Macintosh. We can also transmit bit images or JEDEC files to an onboard BP device programmer as required.

Finally, the documentation of the entire system is handled by an OrCAD hierarchical file structure. This allows me to home in quickly on a needed schematic file, edit it, add notes, generate printed circuit artwork for emailing or faxing to a fab house, or whatever—all from the console. I also generate an occasional hardcopy of the whole system for easy reference, and transmit occasional file backups to the base office for security in case of a disk crash!

The BEHEMOTH project was made possible by good friends who believed in the whole crazy dream and spent many hours helping in the Bikelab at Sun Microsystems. Surrounding Steve, clockwise from left: Maggie Victor, David (Zonker) Harris, Steve desJardins, Steve Sergeant, Greg Davis, and Martin Rowland.

THE BIKE ITSELF

Underneath all the whiz-bang gizmology, I sometimes forget, rolls a bicycle. It's not a normal bicycle, to be sure, but it has all the same requirements as any other bike with the added challenge of handling a gross vehicle weight of about a quarter-ton. This complicates everything…

The recumbent itself is based on a custom frame built by Jack Trumbull (Franklin Frames) of Columbus, Ohio. It is not a commercial product, but should be.

One of the most obvious problems is gearing – with this kind of load, the standard 18 speed setup is useless. *BEHEMOTH* has 105 speeds, ranging from 7.9 to 122 gear-inches. This was accomplished by retrofitting a third derailleur to the tandem-like crossover drive assembly on the left side of the bike.

A 7.9 inch gear with a comfortable pedal cadence of 60 RPM yields about 1.2 mph. While this is necessary for getting a heavy bikeasaurus up steep hills, it does introduce the problem of balance. To address this, we added a set of pneumatically deployed landing gear that can be used under low-speed conditions—or as a kickstand for parking. ("Points for style!" observed a cyclist in Wisconsin as I cruised to a stop with the gear whispering to their extended position, my feet still locked into the pedals.) Just don't call them "training wheels"…

The rest of the gearing range yields more traditional speeds, and I find that a comfortable cadence on level ground yields 10-12 mph depending on my attitude and what I had for breakfast. With a bicycle this heavy, achieving the correct impedance

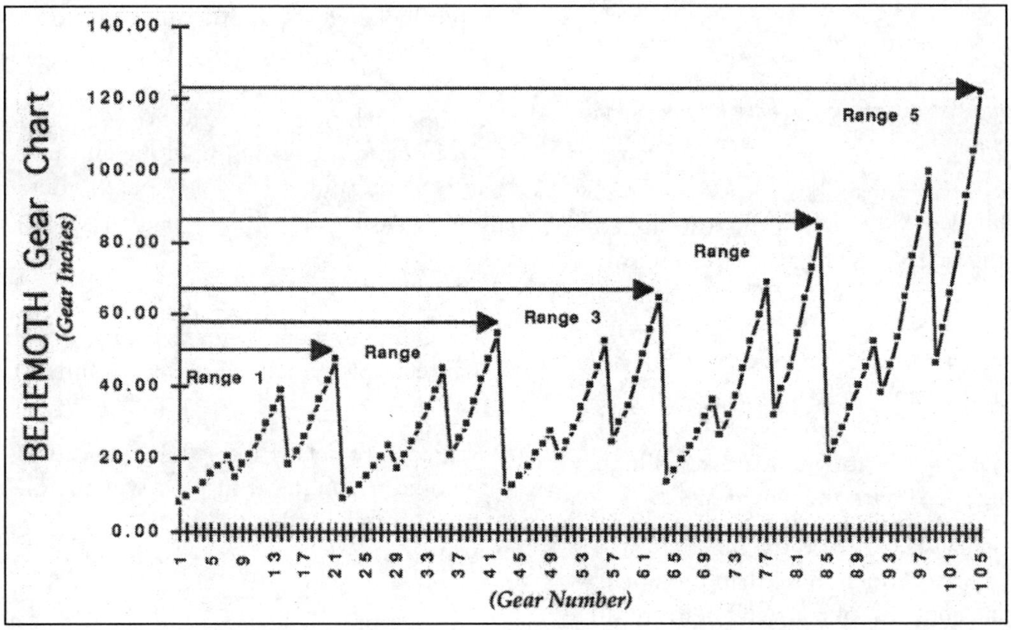

BEHEMOTH's 105 speed transmission consists of two chains and three derailleurs. Think of it as a 21 speed wide-range transmission with a five speed input shaft at the crossover. The pedal drive cog D is 24-tooth; the transfer cluster X is 18-26-34-40-46; the triple T is 19-36-44; and the cluster C at the rear wheel is 13-15-17-20-24-30-34. By the calculation (D/X)(T/C)*27 this yields a granny gear of 7.9 inches and a tall gear of 122 (this arcane notation can be traced to the diameter of the front wheel of a penny-farthing high-wheeler with the corresponding drive ratio). Naturally, there are not 105 discrete steps in a smooth line—as the above chart shows, the major and minor ranges overlap considerably. Also, as with any derailleur, certain combinations (large-to-large or small-to-small) are pathological. Still, the raw 105 speed system satisfies three critical needs: an ultra-granny hill-climbing gear without sacrificing the high range, a large selection of midrange ratios to optimize impedance matching between my body and a very heavy bike, and a number of "tracks" through the gearchart to allow anticipation of down-shifting needs without awkward multiple shifts under heavy cranking load. Psychologically, this takes the shape of a three dimensional 5x3x7 matrix and isn't as complicated as it sounds.*

The Bike Itself

match is critical, and we have no shortage of gear ratios from which to choose.

Stopping power is every bit as critical as gearing, and the system is designed to carry four separate braking systems...

- A pair of Magura hydraulic brakes controlled by hand levers to take care of "normal" braking requirements.
- A hydraulic disc brake on the rear wheel allows speed regulation on downhills without rim heating effects.
- A mechanical parking brake is immensely useful when stopped—and also as an emergency backup.
- Unfortunately never completed but eminently desirable, another pair of hydraulics on the trailer wheels to be automatically actuated by hitch compression.

(In the "never completed" category, I might as well include the variable-reluctance regenerative braking system that was to have become the front-wheel hub.)

Steering has always been accomplished by a single lateral bar under the seat, linked by a coupling rod and a pair of rod-ends to the front fork crown. With the change from the *Winnebiko II* to *BEHEMOTH*, we modified the bars to allow the hands to rest in a more natural ergonomic position for typing... tests with the chord keyboard revealed significantly less stress and higher typing speed with the new design.

These handlebars (and the nearby seat frame) carry much more than the normal bicycle controls, of course. There are three shift levers, the parking brake lever, hydraulic brake levers, and seven keyboard buttons on each side. The right thumb has a button for the twin air horns; the left has push-to-talk.

Safety is a primary concern, as evidenced by the mega-air-horns and their pressure tank (shared by the landing gear deployment system). Lighting is critical as well—there are two separate beams, spot and flood, mounted on the helmet (Night-Sun). These put light where I'm looking, not just where the bike happens to be pointing. There's also a 4″ sealed beam quartz halogen headlight, a 7″ yellow flasher behind the seat, a red LED taillight on the RUMP, a strobe on the antenna mast, and high-brightness LED taillights and turn signals on the trailer. Various reflectors help ensure that the bike is easily visible from all angles, and there are a couple of bright flags just to be sure.

All bearings are sealed, of course, and the rear wheel is a 48-spoke undished unit built around a sealed Phil Wood hub. This intrinsically strong wheel design has never failed me in 17,000 miles—I've never broken a spoke, and only rarely needed minor tweaks to the truing. Some other components have not fared as well in the past, specifically the headset bearing, which on the *Winnebiko II* typically lasted about 500 miles due to hammering by the unsprung console load. On the new system, I not only switched to the much more robust Chris King headset, but also shock-mounted the entire console on rubber Lord mounts to filter out the worst of the high-frequency vertical acceleration.

In short, the bike carries just enough technology to compensate for the weight of all that technology... but I prefer to think of it as art!

Note: More details about some aspects of the bike are online at http://www.microship.com/bike/index.html.

CAMPING AND LIFE-SUPPORT EQUIPMENT

Extreme technological self-reliance conjures misleading images of immunity from all evils—a sort of James Bond lifestyle of instant electronic solutions to any problem that may arise. Unfortunately, none of the gizmology protects me from rain, cold, darkness, dehydration, hunger, overheating, loneliness, exhaustion, pain, injury, and all those other threats associated with being an intrinsically biological hunk o' fragile meat masquerading as the engine of a delicate and insanely expensive vehicle. We're working on the problem…

In the meantime, *BEHEMOTH* carries a full suite of camping and life-support gear, including extensive medical supplies, emergency food stocks, a filtration system for wild water, emergency flares, basic weaponry, camp cooking equipment, harsh-weather clothing, subzero sleeping bag, lightweight street clothes, solar shower, and a lot of other things that make shopping at places like REI so addicting. It has all paid off at one time or another.

Tent choice is critical, largely because I don't like leaving the bike exposed to condensation and human view overnight (or when I'm away from camp). The solution is a huge free-standing Eureka Equinox 6 dome, providing not only weather and tamper-protection for the bike but also a spacious living and working environment.

A secondary reason for the porta-condo, by the way, is the immense difficulty of dragging this whole machine into a motel room. Often, "credit card camping" is the most desirable option, especially when a real shower is starting to sound supernaturally appealing or when conditions do not invite camping. In the past, I have always managed, but it can be a frustrating and difficult end to a riding day. With the porta-condo to keep the glittering gizmology out of sight and a dependable security system to alert me to intrusion, I can actually leave it in a standard parking space… though I prefer not to. A tiny wheelchair wheel mounted horizontally on the trailer frame lets me tip it sideways to drag it through narrow motel-room doors.

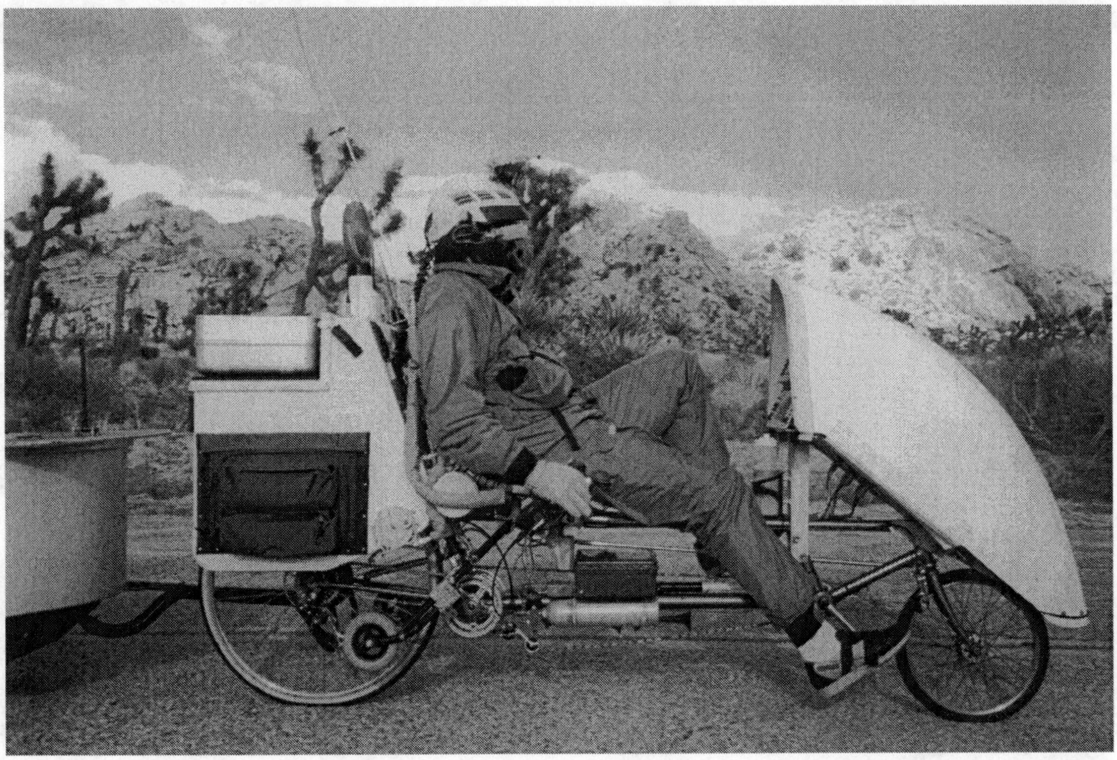

BASE-OFFICE AND NETWORK SUPPORT

It doesn't take long to discover, once you hit the road, that no amount of on-board technology can satisfy ALL the needs of a nomadic business. Without going into exhaustive detail, I can report that this whole affair would be impossible without two essential support facilities: online networks and a base office.

The online world—which I long ago dubbed Dataspace, although history has favored Gibson's *Cyberspace*—has been my home for years. It is a multilayered place that is every place, yet no place at all; it is a new world whose national boundaries are etched along corporate and intellectual lines instead of geographic ones. The Internet *(augmented heavily back in the early '90s by now-superfluous commercial services and various bulletin board systems)* comprises my electronic neighborhood and essential conduit for business information traffic. Without the Net, quite simply, this adventure could not exist… and my physical location would make a difference.

The other essential component is the home base. A time may come when this is unnecessary, but a number of requirements have yet to be met. Most importantly, much of the world still depends on paper, and the base office serves as a gateway between that physical information medium and the infinitely more portable electronic forms. In addition, such institutions as banks and insurance companies are still more comfortable with people who have physical addresses, and the base office presents the illusion of stability. And, of course, there is the presence of an adaptive intelligent biological information system to deal with the myriad unexpected problems of any business. Automating *that* is quite a few years off.

The Final Tour

So. What happened with *BEHEMOTH?* As you can see from the timeline back in Chapter One, there is quite a disparity between the development-to-adventure ratios of the three bike versions. The difference, I have come to realize, is partly due to the *Roberts' Law of Applied Mobile Gizmology:*

> *"If you take an infinite number of very light things and put them together, they become infinitely heavy."*

It's also due to a more subtle problem: I never took the atlas to the bathroom during three years of *BEHEMOTH* development in Silicon Valley (the last half of which was in a lab sponsored by Sun Microsystems).

Back in '83, when all this began, the underlying motive was adventure and romance. The atlas was my bathroom reading; I pored obsessively over bicycle-touring books. The bike was an essential tool, a streamlined and simple system for accomplishing one basic need: taking an office on the road during an extended bicycle tour. Ten thousand miles of magic and adventure proved the success of this approach, and it's detailed in my book *Computing Across America,* available from Nomadic Research Labs.

The *Winnebiko II* was similar in spirit, though much more complex in implementation. The road fantasy still prevailed—with the new twist of sharing it with my girlfriend. Six thousand miles on both coasts once again demonstrated that our hearts were in the right place. (Read all about it in *Miles with Maggie,* excerpted on our website.)

But an odd thing happened during the *BEHEMOTH* project… the road moved to the background. The bike itself became the obsession—a techno-fantasy, a standalone statement. Not once during the entire project did I take the atlas to the bathroom and lose myself in dreams of the road… I was busy studying trade journals and spec sheets. And I never noticed until I was once again Out There… sweating profusely in the Iowa sun with my 580-pound load.

A radical recalibration of philosophy occurred during that summer of 1991. While I did have a few notable adventures in that 1,000 mile jaunt through Iowa, Illinois, Wisconsin, and Michigan, it became clear that sleek lightweight human-scale practicality had taken a back seat to interesting system architectures and the obsessive desire to integrate everything into one machine (a phenomenon now known around here as *The BEHEMOTH Effect*).

As I was pedaling along the Lake Michigan shore, trying to figure out how to deal with this problem, a strange thing happened. I started gazing out over the water, imagining the bike as a ship. No hills. No crazy drivers (or at least, not too many). Quiet when I want it. Open-ended wandering. Beaches. Hmmm.

Thus preconditioned, and entertaining various aquatic adventure fantasies, I met Christina during Spring '92 Interop in Washington, DC. This jazz singer and unix system administrator had what she openly called her "man trap"—a pair of sea kayaks. A few weeks later we had a rendezvous on a lake in the Adirondacks… and the moment I slipped into one of the tiny boats and glided across the water under paddle power I knew my life was about to change…

Chapter Three

A Passionate Interlude

Dangerous Influences

Maybe it's the music, classic Pink Floyd penetrating me as I write this. Wordless memories overtake the present, obscuring it, rendering the computer puzzling even while practiced fingers perform their familiar little dance. Perhaps madness lurks herein: time is inside-out; the swirling vapors of time are suddenly real. Guitars like scalpels part the callused years, revealing visions of terrible glorious color overlaid upon freight trains rumbling gritty in the night, memories of adventures and obsessions potent enough to raise gasps and gooseflesh… my first astonished discoveries, 30 years ago, that life is a thing of infinite potential.

I recall suddenly a day up Boulder Canyon, long ago, free-climbing far beyond my skills. The rock, hard and hot against my cheek… my legs, vibrating with the tension of death's leering proximity… and on top of it all, that crazy moment when reality gets lost among a dozen hotly competing alternatives—each convincing, each alluring, each equally fatal if mistaken for the real thing. I grinned into the stone and inched impossibly upward, curiously disconnected, vision overloaded, abruptly FREE…

Yes, freedom. That's what is behind all this: the exhilaration of walking empty-handed away from Somebody Else's Plan and sticking out a thumb, leaning against the superstructure of a drawbridge for a thrumming 3:00 A.M. liftoff, dynamiting a love nest with walls where once were windows, releasing the brakes what-the-hell and flying with a shout down a mountain road, releasing reality and flying wide-eyed into the infinity of psychosensory unknowns… it all tastes of freedom.

Try, PLEASE, to capture this. Reach into your past, before marriage and business, and examine the brief gaps between commitments. Inside those gaps are subtle tears in the fabric—glimpses of wild seductive alternatives to everything you knew at the time… other realities inches away, dancing just out of reach, teasing you with infinite possibilities even as you turned dutifully from one chapter of your destiny to the next.

But did you let it lure you away? Did you leap into the unknown and hitchhike off somewhere, not caring where, yearning for the sweet sense of movement and discovery? Did you shock your family and chase a crazy dream, abandoning years of conditioning to let that spark inside you explode into flame?

And what about NOW? Is the notion of staying ten years in one place depressing… but somehow inevitable? Are you doing exactly what you want to do with your life, not only at this moment but at 9:00 A.M. Monday morning and tonight in bed?

The most delicious freedom comes from venturing beyond the assumptions that other people have made about you. The real prisons are those of expectation: denying the possibilities of your life in order to be what somebody else wants you to be. I've watched brilliance

tarnish, fade, and finally disappear in the murk of a stupid marriage. I've seen those capable of pushing the big envelope waste a lifetime waiting for little ones with paychecks, rationalizing lost time with vague dreams of retirement travel and future ventures. I've seen others, constrained by circumstances or interests to a steady job, discard all leftover energy in a nightly haze of television, alcohol, drugs, religion, or dull routine.

I am not a proselytizer for nomadics—or anything at all, really, other than what's already inside you. There are countless ways to explore that, and my own peculiar choices are obviously not for everybody. But damn it, do you have any idea how much brilliance and wit rots away undeveloped? We need to do away with the numbing influences of this mad age and start developing PASSION. What could you teach others if you applied your skills and insights to whatever you love most?

Could you change the world if given a chance, even if only through a tiny increment in the evolution of intelligence?

Today's assignment: do something that involves risk, learning, awe, passion, courage, invention, insight, or the sweet sparking of another's awareness.

Dangerous Influences

Tidal Passion

While we're on the subject, let's talk about passion. It's a driving theme of nomadness, of learning, of life in genera — it's the crystallization of dreams, the lust for evolution, the antithesis of comfort.

Without passion, life is spent waiting... waiting... waiting for someone else to make it all seem worthwhile.

With it, growth is a way of life and YOU are in control.

Passion is not an intellectual notion, nor a psychological abstraction. It often appears for a while in association with sex, but that's not what it's all about either. Passion is raw and all-consuming, and can't be replaced with religion, New Age interpretations of experience, academic compartmentalizations of the universe, pleasure seeking, or a romp up the career ladder. It's intense, almost violent; it renders everything else in life unimportant while driving you on a quest of personally epic proportion.

Something like that is not to be taken lightly, especially if you once felt it but now sense it slipping away.

The problem is that our cultures, in different ways, discourage passion — although not overtly, of course. We're politely encouraged to excel, to invent, to make something of ourselves. But the people who actually do so have had to struggle past the boundaries of a society that offers up numbing entertainment, reduces education to the level of homogenization, discourages personal risk in its corporate world, applauds conformity, treats the exceptional as aberrations, and rewards the *successful* with that spectacularly sanitized mediocrity known cynically as suburban bliss.

There's an abrupt boundary between the haves and the have nots, as far as passion is concerned. You can't just dabble in passion — it's all or nothing. Suddenly finding it makes you resent Christians for appropriating that otherwise useful term "born again"; losing it makes you feel dead.

No, there's no such thing as a passion dilettante. Your life is either driven by a grand, magnificent, all-encompassing design... or it isn't.

What IS possible, unfortunately, is to live passionately for a few years then suffer through the agonizing process of watching it slip away — without even knowing whether it's recoverable. It must be a bit like Parkinson's Disease... the mind goes, but slowly enough that you witness your own dissolution and understand perfectly well what it means.

I am discovering, however, that passion can be viewed as a tidal, and thus cyclic, phenomenon. It has been in my life, certainly, with every ebb a slow tragedy and every flow an exuberant celebration of new growth. I recoil from stasis with the fire of a new project... then burn out and fall back into stasis. The question is: how can one short-circuit this process and keep passion ALIVE? Could we survive nonstop passion, day in and day out? Is endless passion even possible? If we see it slipping, can we snatch it back?

One way, I think, is with landmarks. For me, it's a strange mix of favorite road music, an amusing juxtaposition of nomadic system design concepts, fantasies of magical encounters Out There, and a few freeze-frame images of intense romance or adventure etched like lightning flashes on my brain.

Another way to hang on to it is by spending time with passionate people—other mad, driven souls who brave the chortlings of the complacent, celebrate risk, and fear not the specter of bankruptcy. It's powerfully reinforcing stuff, and when you forget your own passion, a spark from someone else's can reignite the blaze.

"...there's no such thing as a passion dilettante. Your life is either driven by a grand, magnificent, all-encompassing design... or it isn't."

Yet another way is through obsessive learning: peeking under rocks, exploring different cultures, chasing seductive unknowns, and emerging into the sunlight from the mine of your own specialties to exchange information with those in other mines (a process better known as consulting). Learning is a delicious addiction, even though schools usually present it as a method of working for approval rather than daring to reveal the terrible secret that education is actually a magnificent form of play. Satisfying your curiosity at every opportunity is a good way to keep your passion alive.

Now let's list a few methods that DON'T work:

- Making lists of things to do, especially if they represent the intellectualization of something about which you were once passionate.
- Perennially reshuffling your workspace, filing systems, business structure, software choices, circle of friends, or hometown—all in the name of correcting problems that are interfering with your pursuit of *The Big Dream.*
- Waiting for someone else to come along and solve your problems, or, if you're wealthy, attempting to subcontract your quest.
- Praying, drinking, getting stoned, swilling coffee, playing computer games, or otherwise engaging in any numbing and time-consuming ritual that by direct effect or superstition is somehow involved with soothing your psyche or warding off danger. (Not that all these things are necessarily BAD, mind you, they just don't have anything to do with passion... even though some of them feel pretty good. Why, one day on a coffee buzz I broke two million in Crystal Quest and celebrated with a drink.)

The important thing is recognizing when your passion is slipping—and stopping it before it's too late. The trappings and rewards of past brilliance echo sweetly with the magic of days gone by, and it's blissful to sail upon remembered waves if you ignore the fact that you're not on a boat anymore.

Tidal Passion

Remember why you are. Life is only once, and slips by so smoothly that you can get away with coasting through a whole career and still look pretty good. Think about what you really want. Grasp it with unshakable passion and focused desire. Everything else is secondary.

The Many Flavors of Water

In the midst of this insanely complex Microship project, it's occasionally necessary to have a bit of TOW—Time On Water. This is essential therapy that revives the mind after weeks of mind-numbing daily marathons under fluorescent lights... and reminds us why we're going to all this trouble in the first place. And so, in that spirit, I offer you this little tale as respite from the technical details that fill these pages. The opening story captures the precise moment at which my vaguely defined notions of aquatic technomadness first snapped into focus, inspired by a single magic night on the water.

The Adirondacks... First Touch

Late. Dark. Loons echo otherworldly laughter from afar, their voices following countless acoustic paths to arrive at my ear as if from the depths of space, the real distance impossible to determine. The moon, almost full, filters through deep woods... little patches of flickering motion adding unsteadiness to our walk. The mirror-calm waters of Fourth Lake in the Fulton Chain reflect distant shorelight, starlight, moonlight, and the subtle glow of a solitary cloud, all perfect in concert like the visual equivalent of a distant impressionist nocturne. Claire de Loon...

Every overused romantic cliche of idyllic moonlit evenings was created for this moment.

Christina and I pick our way through the trees to black water, and quietly slip into our kayaks —the sensation more one of *putting on* rather than *climbing aboard* the tiny boats. We adjust to a new reality: afloat in tight personal craft responsive to our very breath, legs disappearing below deck, feet on rudder pedals, hands gripping cold fiberglass paddles, rumps a few inches below waterline. Almost silently, with only the swish and drip of paddles to reveal our movement, the dock falls away.

Surreptitiously we glide into the lake, into the darkness, into the exhiliration of adrenalin-intensified NIGHT. My bow wave is a steady vee; paddles flash white in the moonlight,

rhythmic, my hands cranking smoothly like pedaling feet, the rotation of the feathered shaft already automatic, satisfaction in the silence of efficient motion. Aimless yet swift, we disappear onto the dark watertop, drifting in parallel like twin spacecraft, silent, vigilant, coldly alive and alert in the impossible vastness of infinity. Our trajectories merge slowly, trimmed by microinches of pedal movement.

I look over at shadowy Christina, a blonde ninja of the night gliding beside me in beautiful deadly silence, her subtle wake a phosphor line barely visible yet somehow unmistakable. I murmur sweetness and she laughs softly, the sound amplified by solitude. Our craft collide gently and we lay paddles across each others gunwales to merge them... drifting as one, slowing. Jupiter is brilliant. Atop the dark cold Adirondack water at 2:00 A.M., our humanity is a precious shared event of the here and now, the online world of computerized frenzy far from our thoughts.

The boats tip as we lean together and kiss—lips soft and yielding in the depths of deep space. Sweet. Moonlight in eyes, a smile, Gran Marnier on our breath, the night perfect. With a giggle we push the boats apart and sprint away, circling and drifting, playing commando kayaker, bumping in the night, dancing with moonbeams, hiding in the deep shadows of shore, frolicking like dolphins until the night grows cold and cabin lights beckon...

The Quaking Bog

The next day. Oh my. I could base a lifestyle on this, I observe to myself with an internal smirk as sun-warmed shoulders bend again to their rhythmic task, the miles passing easily, the day sparkling within and without. (The last time I made that prophetic lifestyle comment was in the spring of 1983, while pedaling a strange recumbent bicycle along Ohio farm roads and dreaming of techno-madic possibilities…)

Perfect, perfect. The word keeps dropping into conversation as it does on summer after-noons in the Rockies. Slipping along wooded shores, at one with the water, the wind light and pleasant. Gliding through narrow inlets and creeks, prowling coves, whispering through milfoil and arrowhead, nudging aside the yellow buds of waterlilies, pointing out deer on shore, startling a quacking couple shepherding a busy flotilla of downy yellow-brown ducklings. The muscles of my arms and back growing into a new role, their soreness strangely pleasant; learning the rhythms of power and the behavior of the tiny craft, smiling when a maneuver drops me perfectly alongside my friend to share a moment—or lands me with a whispering bump on precisely the intended spot.

The highlight of this day (actually, a meta-highlight, given that the day itself is one) is the bog on Third Lake. "Marsh," I thoughtlessly label it, unmoved, muscling back toward open water after a quick pass through a tree-lined cove. "No, no, wait," Christina calls, "I want to check this out." Reluctantly, I steer hard to the right and come about for a quick pass… and find myself in another world.

Covering many acres, this is a quaking bog—a floating ecosystem of cranberries, lichens, sphagnum moss, cattails, orchids, redwing blackbirds, and carnivorous pitcher and sundew plants. Awestruck, we creep softly into a long crack barely wide enough for the kayaks, pulling up next to a beaver house of stacked driftwood. We climb gingerly out of the boats—I onto a small island, Christina onto the larger mass… and it's fluid! She bounces up and down, and over a hundred square feet of surface rocks with her; I try the same trick and my island sinks slightly, covering my ankles… then it calves, drifting a foot or so away from the mainland. I pull back with the paddle, feeling like a penguin on an ice floe.

How many years has this been growing? Did it begin with a windblown scrap of sphagnum or a bird-dropped DNA ambassador from afar? I glance across the lake to well manicured lawns and moored powerboats, and suffer a wave of sadness… this micro-world may one day be developed into oblivion as have so many of its diverse brethren. But for now it is a timeless wonder, a living laboratory, a showcase of natural magic.

We return to the water, somehow changed. An idea, already germinated, is beginning to take root.

Whispering Through the Slushbergs

Winter in Wisconsin. It's a cozy morning, but my kayaking clothes are wet from yesterday's crunch-gliding adventure among gently bobbing slushbergs, the GPS tracking distance and bearing back to my host's house as I whisper past an icebound shoreline reminiscent of Carlsbad Caverns (where I walked the twisted subterranean paths of Middle Earth

The Many Flavors of Water

THE MANY FLAVORS OF WATER

only six weeks and 5,000 miles ago). White birds ghostly against white sky. Ice clunking under the hull and tinkling off the rudder. A pair of bemused Irish Setters regarding me from their backyard. The obligatory kayak-mobile contact on two-meter ham radio. The irrational temptation to turn east and paddle the 80 miles to Michigan...

The Epic Crossing of Morro Bay

Morro is a gentle bay, protected from the raging winter surf of the not-so-very-Pacific by a long finger of dunes. West of these, waves thunder and hurl a haze of suspended spray into the chill air; east lies the quiet bay, only an hour or two by kayak from any point to any other. (Unless, of course, you get caught in tidal currents—a few weeks earlier on a solo trek to Morro Rock, it took a serious death march to make it back before dark.)

We launch from the little espresso shop in Los Osos, fueled by righteous caffeine and the mild adrenaline of casting oneself upon the water in a tiny craft. A small crowd of latte snatchers watches us, wide-eyed and chattering; we stow the kayak dolly and off we go, savoring as always the freedom of the water, dreaming of Microships and adventure.

It doesn't take long at all, however, to discover that Morro Bay is a shallow body—for most of the mile-plus crossing we can clearly see underwater grasses gently leaning with the tidal outflow (and I can report that when you see birds standing on the water, you should take that as a clue that it's NOT a good place to paddle). But we press on and beach the boats on the dunes to clamber across their desolate landscape, marveling like children at the strange shapes wind-sculpted into the sand by recent storms, drawing hiragana with a stick, and generally dropping into first-to-set-foot-on-a-new-world fantasy mode. I keep glancing at my watch, wondering when it would be wise to head back.

About an hour after we should, we do. The folly is immediately evident, and as the water drains away we find ourselves seeking little rivulets in the vast expanse of mud that might get us closer to the shore before we have to start slogging. Like the veins of a leaf are the drainage patterns, and on a distant capillary, a hundred meters or so from the espresso shop, we can float no longer.

To the knowing smirks of tide-conscious locals, we plunge into knee-deep black stinky mud, barefoot to prevent sandal-sacrifice, and fight our way slowly to shore. When at last we arrive (flicking off one leech), someone tries to sell me a set of tide tables. The learning curve of the aquatic life inches slightly higher...

The Many Flavors of Water

West Seattle, Alki Point. I spend my days in a tiny beachfront apartment: watching the tides come and go, and with it, the slow turnover of beach detritus bringing messages from afar: fishing floats, beer cans, tennis balls, seaweed, a hardhat, worn glass, rubber gloves, respirators, sawlogs, signs, and a giant styrofoam block that was in a different spot every morning until I launched it back across the sound last night in an offshore wind. Civili-zation's debris rendered vaguely romantic by the actions of wave and tide...

At night I clamber the breakwater—a ragged jetty of treacherous boulders sharp with barnacles at low tide, slick in places with dark slimy patches of seaweed. Trusting my life to floppy sandals, I pick my way seaward in the blackness to my favorite perch, where I recline, slip into reverie and watch the lights: nearby apartments, distant shores, massive ferries, sweeping high beams on shore, buoys and beacons, tiny twinkles of stars, the busy air corridor to SeaTac. Last night the moon was full... making a fine sheen on the back of a harbor seal annoyed by my clumsy appearance on his home surf.

It's the water that lured me here, and it's the water that's out there now, steel gray under Seattle skies, marked momentarily by a freighter wake, random pockets of glassy calm, a blue heron and the gaggle of seagulls

The Many Flavors of Water

who steal his sushi-on-the-fin, and a few heads of kelp dotting gentle chop. Further out, the islands: Vashon, Blake, Bainbridge. I have kayaked to the latter two, miles of slow paddling, accepting the pace, adjusting to the two-dimensional world of water after years of one-dimensional roads, looking up occasionally in unconscious extrapolation and seeing flight as merely another degree of freedom.

I seem to be a perpetual quest for water.

Water. It's endlessly alluring—and it touches me the same way the original road-fantasy once did, but more so. Not only is it symbolic of freedom, but, at least in nice weather, it's intrinsically pleasant (unlike asphalt). And what has fascinated me recently is that it contains infinite variety: everything about it changes from place to place, from moment to moment. The myriad interactions of current, tide, and wind. The violence of storms, and the dramatic effect of distant ones. Clapotis and williwaw, chop and swell, breaker and rip, eddy and bore. The smell and taste, color and density. The wildlife... and the human company too, ranging from crusty fishermen to buzzing jet skis, from drunken power boaters to massive freighters, from gentle creatures smiling from shore to the laughter of children carrying across the watertop. The surrounding land, the air, the FEEL—all alive and evolving in real time, changing from placid to terrifying with the wake of a too-close ferry, startling me with interacting wave reflections from a stone seawall, delighting me with the sight of an osprey nest or frolicking harbor seals, saddening me with visible pollution viewed in close-up, eliciting <pangs> with the sea-view of a flesh-strewn beach, and offering a slow revue of architectural styles ranging from stained industrial ugliness to the manicured riverfront estates of Old Money. All that and more—yet without the killer hills, wheel-eating road surfaces, impatient drivers, and broken glass that have been an integral component of my nomadness since 1983. It does have its life threatening components, of course, but they are an attraction, in a perverse sense: grand, impartial terrors of the deep as opposed to the pathetic but no less painful demise lurking behind every swerving pickup truck.

"Hey, this is fun; I think I'll base my lifestyle on it for the next few years." Sounds impulsive, but how could I do otherwise when the tools exist to drift quietly and efficiently on sun-sparkled water, sitting so low that I feel a part of it, yet linked better than ever before with global information networks, friends all over, and my own on-board network of alluring machines?

Microship Substrate Development

Microship Substrate Development

Chapter Four

The Microship project began with almost primal simplicity in 1991 as I pedaled north along Lake Michigan in eastern Wisconsin... if I could have wrapped a fiberglass hull around *BEHEMOTH* and pedaled 80 miles across the lake instead of huffing all the way up and around, I would have done so on the spot. But the bike was too heavy for such silly ideas, not to mention my on-board suite of non-seaworthy electronics that would, speaking optimistically, last about two days in a nautical environment.

Shortly thereafter, while wandering the US in the Mothership on a speaking tour (doing trade show and corporate gigs with the bike on stage), sea kayaking at every opportunity gave my yet unfocused water-fantasies shape and direction. In early 1992, I announced the new project on the Net, dubbing the computerized kayak *LEVIATHAN* to echo the acronymic moniker of my bike. Over the next year, as I continued the tour, schlepping *BEHEMOTH* between speaking gigs and TV appearances, my thoughts were elsewhere... layering communication and control systems onto a kayak... or gee, maybe a catamaran built of two... or gosh, possibly even a trimaran built of three.

Hmmm....

I had no clue at the time that I was about to undertake eight dedicated and expensive years of system design and redesign, fiberglass fabrication, changes of fundamental direction, establishing a succession of labs up and down the West Coast, developing and maintaining more than 150 sponsor relationships, constant questing for volunteers, and the most expensive (and educational) project of my life. It's a good thing I didn't know all this: it surely would have been intimidating enough to squelch the whole project.

A lot of our work here goes into mechanical design —this is one of those insanely complex projects in which electronic component choice affects gross morphology, which in turn defines constraints that affect the choice of electronics. Such psychological race conditions can only be resolved by tweaking the granularity knob until the correct solution congeals in a flash. Put another way, I form a fantasy of the desired results and allow input variables to float until an unanticipated combination satisfies the fantasy-comparator, pretriggering a snapshot of the conditions that yielded it before accumulated pondering-propagation delays introduced conceptual drift. (This is not the design methodology you read about in engineering textbooks, even though it's a good summary of life in industry. Most textbooks promulgate the myth that structured methods and sequential procedures will get you there. But beating trade-offs requires inventing new rules to break, not bumping into the limits imposed by old ones.)

– SKR, on engineering at Nomadic Research Labs

Microship 1.0: Kayak Hacking

By the spring of 1993, after much inspired schmoozing around the country, I had located a development site—the engineering building at University of California's San Diego campus. My early conceptualizations of this vessel (now dubbed the *Microship*) were as ambitious as they were primitive, and the first two years saw a wide range of marine architecture variations, interesting learning curves, and a huge amount of development work on the embedded control and monitoring systems with the help of our students. To address the sprawling list of ambitious design goals, the boat grew... and grew... to a 30-foot amphibious center hull with detachable double-kayak outriggers. With the help of a marine architect fresh from an America's Cup winning project, growing interest from sponsors, first-class visualization tools at the San Diego Supercomputer Center, and a team of engineering students, the Microship concept began taking shape.

I didn't know it yet, of course, but this shape would change many times before converging on a final configuration. The good news is that while the nautical design evolved through various phases, our embedded control network and other on-board electronic systems developed along a fairly consistent path... though ever more seductive technology inevitably raised the spectre of *creeping featuritis* while replacing clunky old methods with the latest browser-based front ends and wireless networking tools. (We'll tell you all about the current system design in Chapter 5, in the meantime sparing you the grim details of life back when RAM was measured in "K" and a 105-Megabyte hard drive was an awesome and unbelievable thing.)

I find it enchanting to take what I've earned the right to call a *Long View* on this project. As I reread the early Internet "Nomadness" postings (of which there have been over 300 since the beginning of my wanderings), I'm fascinated by the extent to which our earliest intuitively correct design concepts turned out to be the best after all—and how far we diverged from the tightly integrated human-scale specification before returning to our roots with the benefit of learning curves traversed enroute.

So let's look back six years to the spring of 1993 and see how all this started:

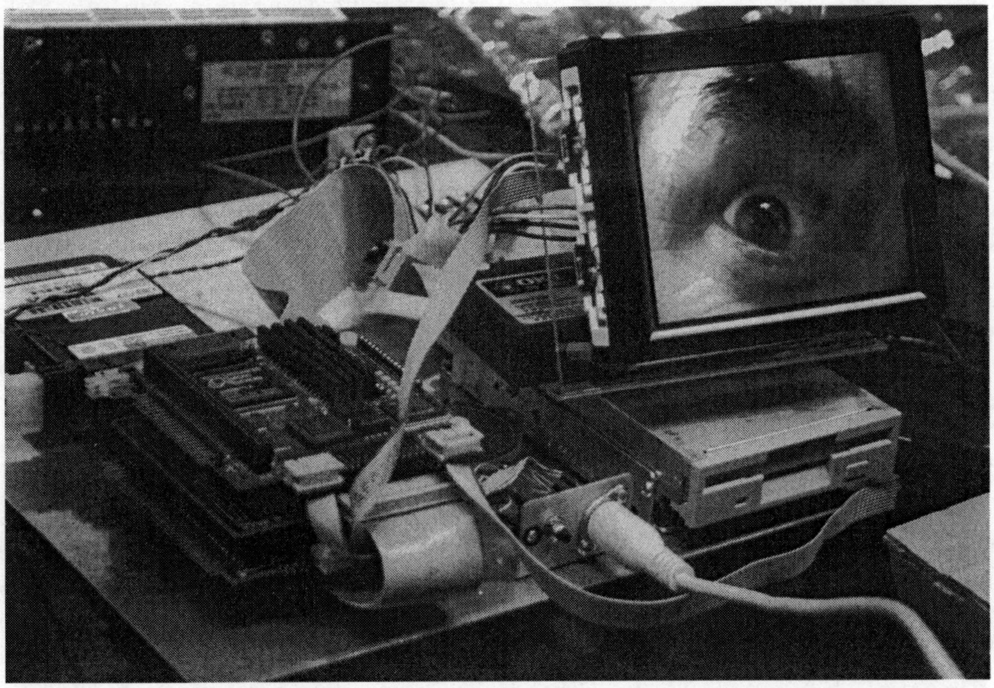

Ampro 286 PC and sharp active-matrix display... first approach to the Microship front end.

FLASHBACK TO SPRING 1993

Well, I have good news! It has been one year since I posted "*LEVIATHAN*: Call for Discussion" to the Technomads mailing list and started tossing around ideas for this new project. A few months later, I began the quest for a host lab, convinced that my next mid-life career change would involve this boat. After a few false starts and a number of interesting offers, the problem has been solved.

I have just arrived in San Diego with nearly everything I own, with the intent of spending the next two years building the new machine at the Scripps Institution of Oceanography and UC San Diego campus. The setting is nearly perfect, and can be described as a high-tech blend of academia and real engineering... conveniently located on an idyllic nude beach with year-round optimal weather (but for an occasional touch of June Gloom).

It's a bit early to report on the nature of relationships and such, but the general scenario is direct involvement with the oceanography community as well as close connection with the engineering departments at UCSD (as a visiting lecturer). The Microship will be a multidisciplinary student project platform, and the affiliation gives sponsors the option of working through education support channels.

The lab itself is not yet precisely defined—more on that next time. As of this date, I have completed the speaking tour, offloaded a truckload of bikelab equipment into what will later become the fiberglass shop, established an office in the engineering building, and am beginning the process of moving into this diverse community.

While all this has been going on, I've been working on Microship design, so I can FINALLY start telling you about it in something other than rhapsodic, abstract terms...

Stated simply, the whole idea of the Microship is to extend the design principles of *BEHEMOTH*—a human/solar-powered mobile workstation designed to render my physical location irrelevant. The new project is an aquatic tool for nomadness, significantly updated to represent the capabilities of 1993-1995 technology. Once the lab is in place, I will work for two years with the development team, then launch in 1995 for a few years of open-ended adventure linked to the network more tightly than ever before.

Much of this connectivity is devoted to the original objective of maintaining stable publishing and personal relationships while traveling full-time. This is no small matter:

Two views of the first CAD model, by Len Wanger, San Diego Supercomputer Center.

most serious travelers report inability to remain current with news and industry developments, poor productivity, loneliness, and homesickness as side-effects of being on the move for years. I have solved these problems by depending on network connections to move my perception of stability into "Dataspace," and the new project will extend this metaphor by radically increasing communication channel bandwidth and the sophistication of front-end software tools.

After exploring a number of configurations and brainstorming with naval architects, kayak designers, sailors, and mechanical engineers, Microship overall design has stabilized into a pedal/solar/sail-propelled trimaran consisting of three sea kayaks—a large, beamy double (about 22´) for the center hull, and two sleek singles (about 17´) for the amas, or outriggers. These are linked by folding solar panel arrays totaling 664 watts (about ⅞ hp) of capacity in full sun, occupying approximately 4´x 8´ on each side plus packaging overhead. If you envision a big double kayak with a sail, joined to a pair of medium kayaks by a couple of solar ping-pong tables, then you have a rough idea of the general plan.

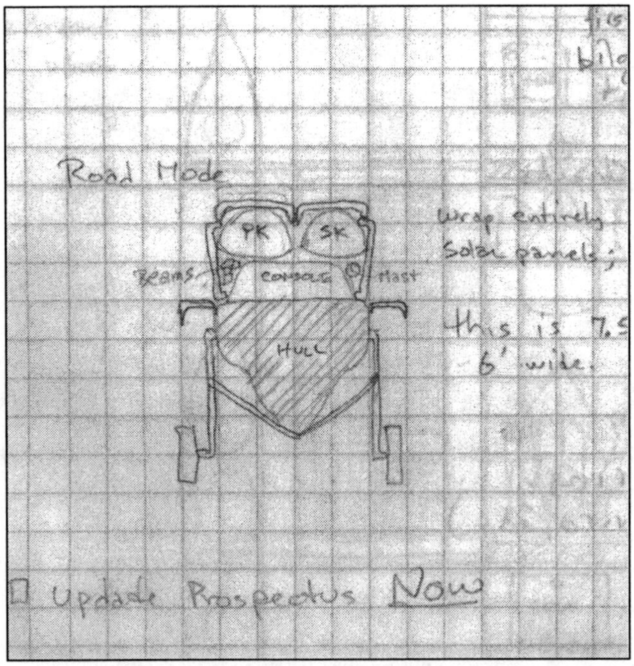

The first conceptual sketches were uncannily close to what rolled out of the lab seven years later... after many sidetrips.

Numerous hatches and sealed equipment bays in all three hulls provide space for packs, comm/nav gear, water processing, food storage, tools, and so on. Given the stresses encountered in rough seas, the composite layup will be considerably stronger than that of a normal lightweight kayak, with special attention to stress transition areas surrounding the attachment points for the stainless steel folding beam assemblies.

The main hull will have a custom deck plan, though the singles should be more or less standard with the exception of outrigger-support fixtures. In front of the main cockpit, a raised console will carry Mac, video, and control system LCDs, as well as the Marine VHF panel and a small cluster of minor controls. The bulk of the electronics will be in a sealed hatch, communicating with the console via the on-board network. Other major deck fixtures include whip antennas, satellite earth station, GPS antenna, a glass video turret, lights, drogue, anchor, compass, and the usual suite of survival and safety features such as grablines and cleats.

Propulsion is based on three power sources. On sunny days, the solar array will provide approximately 50 pounds of thrust from a resting state, yielding cruising speed estimated in the 4-5 knot range via a thruster integrated with the rudder assembly. Power can also be provided by two sets of pedals linked to efficient generators, yielding 100-150 watts per person. While this is about 25% less efficient than a direct mechanical linkage, the weight and complexity are dramatically reduced (and integration with the power management system allows battery recharging using human power under any conditions). Finally, a sail rig can be deployed to allow wind power—the configuration is yet unknown, though good alternatives of appropriate scale are available.

FLASHBACK TO NOVEMBER 1993

That was a pretty tight spec—a micro-trimaran made of kayaks… and it's almost agonizing from this perspective to see how close this was to the design that eventually emerged from the shop… had I only known! Notice in the following how five succeeding months of system design managed to bloat the center hull—from a double kayak up to a 1.5 ton 30 footer (though still with detachable kayak outriggers despite strident warnings from multihull experts that those hulls are the wrong shape). This distortion of the original elegant concept was attributable to the insidious presence of The BEHEMOTH Effect… as the growing list of things that I absolutely needed on board quickly outpaced the original sleek human-scale design. Ah, technolust. Here's a posting from November of '93… note the scary reference to this becoming "the BEHEMOTH of multihulls"!

For the past few months I've been inhaling, so to speak—laying the complex groundwork required to support a complex project. (Here we go again.) At the moment, I have about 40 student volunteers, and the University is providing course credit for participation in many of the more interesting engineering projects. There are many… the Project Catalog details 46 of them, not to mention a growing list of miscellaneous tasks. I'm starting to appreciate the complexities of being a good manager…

First of all, the Microship is not really a kayak anymore—nor a trio of kayaks formed into a trimaran. I've been calling her a "kayacht" lately—with a 30 foot center hull 4´ wide at the waterline, 18´ overall beam, and 1.5 ton estimated weight, the term "kayak" is no longer accurate. We're keeping one critical component, though: the outriggers are detachable pedal and paddle powered double kayaks based on Current Designs "Libra" hulls, now on order. (These amas will be extensively customized with special deck fixtures, bulkheads, wiring, thruster mounts, high-strength couplings to crossarms, submersible cockpits, raised cowlings for pedaling clearance, and deeper-than-normal flotation compartments.)

The center hull is being designed by Nelson/Marek, the San Diego yacht design firm that created the Stars & Stripes catamaran as well as a number of custom yachts and racers. We're having fun with this, pulling out all the stops, ignoring issues of manufacturability and addressing brutal trade-offs that are normally immutable. It's not fully defined yet, but we're dealing with a number of odd concepts that may make this the *BEHEMOTH* of multihulls: kick-up adjustable-depth leeboards for CLR-tuning, floodable ballast chamber to control effective dihedral, free-standing schooner rig, engineering department forward of the cockpit with rackmount pressurized enclosures and fold-down work table, folding solar arrays on polypropylene honeycomb substrates, thruster and pedal drive on each detachable kayak/ama, pop-up crew module for lounging, complete navionics station, and deployable landing gear to turn the collapsed system into a trailer. (To be towed by a gas-burner, not a bicycle!)

None of this exists yet, but we've been working on design concepts (with an actual-size cardboard/lumber mockup of the cockpit segment dominating my lab), gathering composite materials information, working on frame stress analysis, and learning boat-building techniques. No doubt there are surprises in store, but that's part of the appeal… think how boring it would be to just go out and buy a yacht! <wistful sigh>

Incidentally, as we study multihull design an interesting parallel is becoming apparent: multihulls are to monohulls as recumbent bicycles are to diamond frames. They're faster, more interesting, annoying to old-timers, and cover a wide quality range from exquisite to garbage (the latter helping to confirm the negative comments of traditionalists). They both attract wizards and nutcases, leading to odd alliances against Old Methods; they've both been banned from traditional sanctioned races after blowing everyone else off the course. And they are both so undefined that designs have not converged upon a few established standards, but instead show up in ever more radical configurations as designers are drawn inevitably to the challenge of pushing the envelope. In short, both are where the action is.

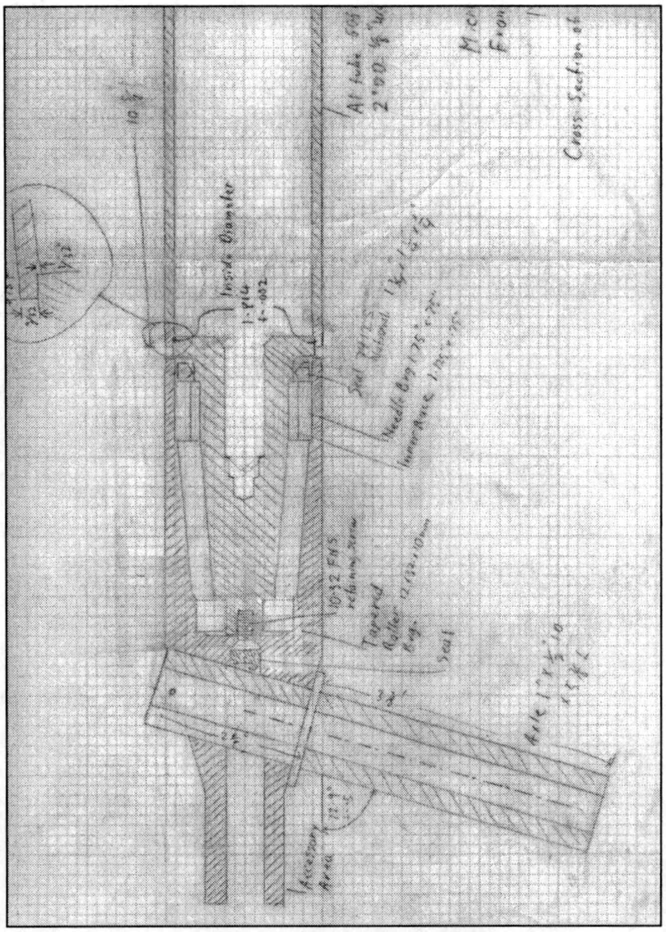

Landing gear design by Bob Stuart.

Microship 2.0: The Fulmar Epoch

That all sounded wonderful, but in April 1994, "downsizing" became the dominant theme. I watched the design bloat further and further as every issue of *Cruising World* and *Ocean Navigator* added must-have items to my wish list of on-board tools, until one afternoon I caught myself spending hours sketching mad images of a huge aquatic motorhome with hydraulically deployed outriggers and pushbutton conversion into a diesel land yacht. That was too much. I recoiled violently, realizing that I could satisfy the primal need for human-scale simplicity while increasing the probability of actually getting on the water in my lifetime by buying and modifying a tiny commercial trimaran, the Fulmar-19. With the boat actually hauled up the outside of the UCSD engineering building into my third-floor lab (a story in itself), the project took an abrupt turn toward practicality… though a dramatic two week adventure with my partner of the epoch through Puget Sound and the San Juans turned up a few fundamental problems with this off-the-shelf solution. Still, this was a grand and amusing interlude in the evolution of the *Microship,* and lay important groundwork for what was to come much later…

Join us now as we take a gonzo, rollicking journey in the Fall of 1994 aboard two matching 19´ pedal/sail trimarans.

The Maiden Voyage

It's all just an abstraction, really, until you actually DO it. For three years I've been reading, planning, designing, learning, and even getting out on a boat now and then. For Faun, it has been even more of a change: we met, found ourselves sharing the same dreams, and decided virtually overnight to go for it. Late nights in the Microship lab and a few sailing classes on placid Mission Bay comprise her background; mine, though more extensive, includes little more in the aquatic reality category but for a handful of coastal day sails and a few dozen kayak trips here and there.

No matter: lack of experience never stopped me before. Full of grand visions of sparkling sunny days and erotic nights in cozy island campsites, we piloted the diesel truck up the coast from San Diego to Seattle in early September — stopping to visit sponsors and friends, standing with them around the trailered boatlet and explaining the dream. "Well, after this little two-week shakedown in the San Juans and Gulf Islands," we would say blithely, "we'll spend a few months integrating the rest of the systems and then truck to Juneau, sail down the Canadian coast, haul them to Edmonton, traverse the North Saskatchewan River across Canada to Lake Winnipeg, hop down to Lake Superior, then cruise the Mississippi, through the Gulf, around Florida, and up the ICW to Maine. Then Europe, of course."

They would nod sagely and offer bits of advice, while we'd glance at each other and try to imagine actually living on a pair of tiny Fulmar-19s. These are trimarans on the scale of kayaks, with center hulls 27" wide and a few wee hatches for gear. I've located a second one for Faun, thus forming the world's largest flotilla of Fulmars (only about ten exist), and we spend much of our time puzzling over packaging, gear choices, and systems integration. Making the jump from pleasant day sails to 10,000 miles of open-ended wandering is quite the extrapolation!

First things first. We arrived in Seattle, borrowed a boat for Faun from Saint Cy, spent two days in frenzied last-minute preparation, and launched for the two-week maiden voyage...

All-nighter: Seattle to Port Townsend

By the time we actually assembled some 300 pounds of packs in our staging area and dealt with the complex logistics of two boats, two trailers, and one truck, it was 1600 on Thursday, September 15. No problem, we'll just sail to some quiet little town like Kingston or Port Ludlow, find a cozy B&B, and then zip on up to Port Townsend in time for my talk at the Sea Kayak Symposium the next day at 1430. Hey, we have 22 hours... piece o' cake.

We set out from Shilshole Marina in the Ballard area of north Seattle with many a sparkling grin and comments along the general line of, "can you believe we're really doing this?". All our senses alive, we noted every little shore feature, ship, and scrap of flotsam. Of course, there was no wind to speak of, so it was three hours of pedaling before we entered Appletree Cove and

The Maiden Voyage

approached the town of Kingston — the sun low in the sky. The plan, simply, was to dock and find food, then decide whether to spend the night or press on.

But the wind, that sly vixen, seductively rippled the water and filled our sails. Rainier's snowy peak glowed orange in the low angle light of sunset. The Edmonds ferry blasted her horn as we sprinted across her bow <gulp>, a HANJIN freighter scudded north in the traffic lane, and distant city lights twinkled crisply atop their rippled reflections. How could we stop now? In no time we had rounded the point and were beating northward, dancing, racing across sparkling water — high on the wind, on each other, on adventure. I pushed my vague misgivings about night sailing into the background and set course for Point No Point.

Three hours later, in darkness, we rounded the point. It had been a long beat into the wind, but the night was yet young and hey, there was a sense of delicious madness afoot. The plan now was to scoot over to Foulweather Bluff, round that, and duck into Port Ludlow around midnight to look for a place to sleep.

For some reason, though, the going became painfully slow as soon as we rounded the rotating white light of the point. We became deeply familiar with the houses on shore, one by one. We would pedal hard to go nowhere; then try tacking out only to return to the same point. Hmmm. After about an hour of this death march, the growing strength of the flood current was painfully obvious: we were like insects trying to claw our way up the faucet that was filling Puget Sound. I kept staring at a point that marked the mouth of Skunk Bay (and its imagined respite) less than a quarter mile ahead, but it became clear we could never make it.

Fortunately, someone had installed a mooring buoy, and I struggled up to it, snagged it with my boathook, and tied up. From this perspective, the current was obvious — raging and bubbling past me like a fast river, bull kelp streaming along the surface, my dropped Petzl headlamp scooting rapidly back to Seattle. Faun worked her way over to me and we rafted, then congratulated ourselves and wolfed down peanut butter and jelly while enjoying the novelty of the craziest midnight we've shared in weeks. Little 2″ jellyfish zipped by, flashing bright green when touched; wee organisms did likewise and turned a sweep of a hand into a mini-constellation. Despite the exhaustion of the struggle, there was magic afoot...

The attempt at sleep was unsuccessful. Slack was scheduled for 0222, and we stretched out on the tramps to wait. But the boats were banging together in a rather distressing fashion (fenders notwithstanding), and water kept working its way between me and my Therma-rest, puddling atop the non-porous fabric of the tramp as I shivered under a wind-whipped space blanket. Another few items for the TO-DO list...

Just before slack, the current down to about a knot, we decided to cast off and continue. Faun untied her boat and drifted down-

current… and kept drifting. A frightened voice from the darkness: "Steve? I need help. The pedal drive is jammed." Kelp, no doubt. As I worked to free myself from the buoy, she sailed toward shore — the composition of which was unknown.

It turned out to be sand, fortunately, and I sailed over (through a kelp bed) to join her. At this point, a serious temptation presented itself: camp on the beach under the trees and deal with it in the morning! However, there was still that speaking gig 20 miles away, so after a brief respite we set off again. Faun was having a few doubts, especially when I pointed toward our ominously dark destination.

Those doubts deepened significantly when the fog rolled over us. Within the hour, at about 0300, we were engulfed by soup, sailing by GPS alone. The blue backlight of the Traxar, red and green spray flying past the navlights on my bow, occasional bioluminescent sparkles from the deep, and the rollicking nav lights of my partner became my world — Double Bluff and Mutiny Bay somewhere ahead. I checked Faun's status via the Motorola GMRS handheld radio: "worried" came the answer in a voice small and unfamiliar. But the fog parted to for a few minutes to reveal a dramatically clear night sky, and sure enough, ahead lay the green beacon of the point and a row of lights edging the bay. We savored the luxury of visual reference points…

But stopping — even if we were lucky enough to find a place — was out of the question, for now it was nearing ebb. It would be insane to sleep through a favorable current only to wake in time for a contrary one, so the decision to press on was unquestioned. This was to be a long reach — about nine miles up Admiralty Inlet. Again the fog… and with the moon setting there wasn't even a hint of skylight to provide differentiation between water and sky. Just thick blackness, the microworld of a tiny boat, and the immediate waves — nothing more.

This leg became a mad dream, a wild ride of sheer exhilaration and absurd risk. Running 7-8 knots on a close reach in favorable current, we trusted our lives to the miracle of GPS. Faun's job was to keep my lights in sight; mine was to maintain the heading that would clear Marrowstone Point while trying not to think about the fact that we were screaming northbound in a shipping lane in the fog. Indelible in my mind is the image of my friend in a sleek matching boat, flying beside me, crashing through waves, soaked and cold, completely dependent upon me and the technologies that provided our flotation, navigation, propulsion, illumination, and communication. This was deliciously mad and totally consistent with the spirit of adventure.

("Adventure," quoth a netpal, "is the result of bad planning." But I prefer to think of it as the whole POINT of planning.)

It seems that the spaces between waypoints are chapters, for each turn, while part of the same overall trek, adds new and unexpected twists to the evolving story. The next chapter was home stretch at last, and given the northeast wind I was eagerly anticipating a nice broad reach on the last 3.9 mile jaunt to Point Wilson, just north of our home for the weekend. At dawn we rounded the horn that had been teasing us for two miles, turned eagerly west, and… and… so did the wind. Thick white fog and a headwind. Lovely. <sigh>

Slog on. For three more hours we kept at it, cranking our throbbing legs, hallucinating land and ships, numbly alert to the horns and rumbles of freighters and ferries around us. It seems so tame now, looking out from the cliff on this clear sunny day in Port Townsend, but in the soup after an all-nighter it's sinister and exhausting. Faun had to laugh at me: startled by the too-close chest-vibrating foghorn of a ferry, I responded by blowing my whistle – a contrast too silly to contemplate. But they weren't monitoring VHF channel 16 and I had to try something… moments later the monster materialized not 100 yards away,

THE MAIDEN VOYAGE

growled implacably past, then dematerialized so quickly that it seemed but another hallucination of fog demons.

I steered for the Port Townsend horn to take advantage of a close reach, then turned north for the final couple of miles, but even THAT was to be a challenge. Not only were we proving that wind always originates from one's destination, but the currents had returned with a vengeance: it was 1030 when land finally ghosted into focus. My exhausted brain called out to Faun, "there it is! See those big buildings up on the cliff? That's Fort Worden!" The "buildings" were actually a line of driftwood resting on the beach at high tideline, but I was on the right track... a few minutes of drifting easily landed us in a cluster of unencumbered sea kayaks. My first thought: "What? These wimps came here by CAR???"

Distance made good in those 16 hours: 39 nautical miles, 30 as the crow flies, about 60 if you count all the tacking. We do like to jump in with both feet, eh?

This update comes to you from the second day of the Sea Kayak Symposium. I somehow survived my somnambulatory speech about adding electronics to small boats and am learning from others; we're sore-legged but recovering; the Fulmars are presumably still at their moorings (yes, I'm paranoid about security); and we launch tomorrow for Port Angeles — our jumping-off place for the trek across the Strait of Juan de Fuca to Victoria.

The John Wayne Marina

Well, we did indeed launch for Port Angeles, but didn't get that far in one day, or even two. Actually, we didn't get there at all — that's part of the reality of sailing.

By the time the currents allowed us to move significantly west from Port Townsend, it was getting late... so we gave in to temptation and made for Sequim Bay. Now, on the chart, this looked inviting — a quiet little corner between Discovery Bay and Dungeness. It was dark by the time we slogged into the mouth of the bay (oddly, both ebb and flood fought us on the same compass heading). I can't emphasize enough just how demoralizing it is to pedal into current without the aid of sails: it's much worse than bicycling into a head-

wind, where you at least have solid ground to push against. Here, you push against something that is moving the other way...

Exhausted, we poked in the dark around Washington Harbor, expecting from the name to find, well, something vaguely akin to a HARBOR. I was already soaked from the waist down from an attempted beach landing on a ragged hell of rocks and shells, and we found ourselves peering into blackness with inadequate flashlights, looking for anything to tie the boats to. On a hint from a bemused local, we pushed one more mile to the John Wayne Marina.

Now, marinas are not made for people like us. Most boats that patronize these places are yachts, and either accommodate sleeping on board or discharge arriving yachties into their cars. But we staggered onto the dock asking weird questions about tying up two baby trimarans ("that one is my dinghy, you see—can we just pay for one berth?") and pitching a tent, and finally were allowed to camp with the warning that we had to be out by daybreak. And while this did solve the evening's problem, it also demonstrated that marinas are not necessarily the place for tiny little boats.

Shipwrecked on Dungeness Spit

Morning. Dead air again, but in good spirits we pedaled north, making for the tip of Dungeness Spit, slowing to enjoy the shallows, taking a break on a beach, gently raising the idea that maybe we need a larger boat that can accommodate sleeping. Dungeness would be the turning point at which we would either launch across the Strait or run west for Port Angeles, but we spent most of the day getting there.

"Let's swing around the buoy," I casually said as we at last drew near, "there might be some tide rips at the point." As it was getting dark and cold, I turned my attention from sailing to wriggling into Kokatats.

By the time I looked up, the buoy, once dead ahead, was a mile to port and receding fast. We were in a killer flood current pushing us east, and were absolutely unable to find any combination of pedaling and sailing that would yield westward progress. The range to Port Angeles was 15.0 miles, quoth the GPS—and there it stayed, unmoving for over an hour, as we tacked back and forth, trying to avoid losing ground.

At last it relented somewhat and we began clawing our way in rising wind along the outside of the spit. We would tack out into freighter-land and then zoom back into kelpville, beating toward the imagined respite of a state park (Okay, a National Wildlife Refuge) noted in my DeLorme atlas. There had to be a landing, for we could see a dense fog bank closing and the sun was going down. Around here, night sailing in high wind is risky business—the sea is peppered with serious hazards in the form of floating logs and deadheads (logs floating just below the surface, not the tie-dyed kind). Thrice in this windward beat I had to veer to avoid a collision.

The wind was growing, the seas capped with white foam. To our left, rolling waves arriving from the Pacific crashed on the driftwood-strewn spit, a profoundly inhospitable place for boats. But the darkness and fog were closing, and Faun was pulling ahead due to a recent misalignment in my two mast sections (normally we're well-enough matched in sailing performance that we use each other for sail-trim feedback).

I watched in horror as she made straight into the surf zone. A couple of tourists on a sunset stroll stopped to stare. I thought to yell, but she was too far off and the handheld radio batteries were dead... I saw her boat start to bob, then pitch violently... then hit the beach and broach in the surf. One wave swung her broadside, and a moment later she was up to her neck in water, struggling with the boat, trying to drag it up a steep drop through dumping surf. Finally, the gawkers moved to

THE MAIDEN VOYAGE

help and within moments she was — for the moment — safe on the sand.

Great. There was no way she was going to get off that beach under these conditions, and because it was low tide, the boats would have to be hauled some 50 feet into the tangle of driftwood to survive the night. Darkness was closing rapidly, the wind was howling in my ears, and a dense wall of dark orange sunset-backlit fog was bearing down on me. Port Angeles was still over 10 miles away, directly into the wind, and there was no other place to land. Okay, here goes...

I made one final tack to approach orthogonal to the waves, and made straight for the beach, Faun watching wide-eyed. As I felt the surge of the surf, I had idealistic images of riding smoothly onto the sand and hopping out to drag the boat kayak-like up the beach, but it was not to be so simple. The boat hit the beach hard, veered, took a wave that filled the cockpits, washed out, slammed again... somehow I found myself in the surf hauling on the bow painter, aided ineffectually by the strollers who were being careful not to get their feet wet. Every wave slammed the rudder hard to port, bending the yoke; the hulls ground on the rock-strewn beach and gave up precious gelcoat with each painful scrape. I shouted directions to the locals, and somehow we dragged the boat half out of surf – giving me enough time to frantically pull drybags of gear out of flooded hatches.

We finally pumped out and, with four groaning people, hauled the *Microship* and her sister ship up the beach and into the tangle of bleached logs above tideline. By now it was fully dark and the tourists beat a hasty retreat, leaving us with the amusing task of pitching camp in heavy wind and wet sand. Everything was sodden and gritty, and it was nearly 2300 by the time we shivered into our Caribou bags and tried to sleep. I set my alarm for 0345 to monitor the boats through high tide.

Through all this, we were mildly paranoid about getting caught camping in a wildlife refuge, but hey, any port in a storm. As it happened, the ranger was a friendly fellow, somewhat amused by finding two baby trimarans and a tangle of gear on the spit. With more help from strollers, we aimed the boats at the beach in morning light, waiting for a high tide to provide a window of gentler surf that would lift us off. But two strapping lads came by, stripped down to shorts, flexed their muscles for Faun, and launched us straight out to sea — to hell with counting the sets, this looks good, HEAVE! She almost pitchpoled backwards in the next breaking wave, but we made it.

Death March Across the Strait of Wanda

It seemed surreal, all of a sudden, to be afloat, the now-sinister Dungeness Spit slowly receding. Ahead lay Canada, only 20 miles away, and we had happy images of a long skimming reach in the legendary winds of the Strait of Juan de Fuca.

But the water, within the hour, became so glassy that we could read each other's sail numbers in the reflections.

Ah well, deploy the outdrive and pedal. And pedal. And pedal some more.

For ten hours we pedaled, unable to rest without drifting backwards in the current.

It had its moments, of course. An odd optical illusion that looked as if a giant tidal bore was bearing down on us from the entire eastern horizon. A dolphin arcing through the water before my bow. Ham radio autopatch contact with David Green in Victoria, informing him of our ETA. The rich beauty of sunset, mauve and peach blended and swirling gold-speckled through a sea muscular with ocean swell. Moonrise, full and huge behind Faun. The close encounter with a freighter. Chat with the friendly Victoria ham community, concerned for our safety and monitoring our progress. Occasional weather reports on the marine VHF, forecasting calm on the strait. The painfully slow increments of the Traxar

coordinates; calling out a new mile marker every half hour as we fought our way into the current at about two-knots. Slowly nearing lights of the city...

And suddenly, impossibly, the passing of the green light, and then the red one, and dark Victoria harbor with Dave waiting in a dinghy... pedaling that last mile to the customs dock between seaplanes and a megayacht with helicopter perched on the afterdeck. "I feel like I'm in Disneyland," said Faun in a weak, slurred, exhausted voice, gazing past toys of the rich at the gaily lit parliament building. We had just pedaled a half-ton of stuff over 20 miles. After clearing customs by phone (this silly process now reduced even further to absurdity: "are you carrying any weapons? do you have any drugs?"), we were hauled to Dave's house and tucked in.

Realizations in the San Juans

I write now from a friend's cabin on Henry Island, facing Mosquito Pass. It's Saturday dawn, and the stinkpotter exodus is in full swing as the recreational fishermen race out into Haro Strait for the day's kill. Our boats, tiny and frail, bob about on their shared mooring buoy as the motorists zoom past. They remind me of various breeds of dogs: big stupid galumphing ones with fenders gaily slapping the hull, squat vicious ones kicking up absurd wakes, little dainty ones puttering precisely along, lean racing ones flying across the watertop, expensive purebreds parading in coifed statuesque splendor, scatterbreed mongrels colored primer and dirty white. All are annoying to a sailor.

The past few days seem dreamlike... exploring Victoria, getting to know Dave and his family (world travelers fresh from an extended stay in Fiji), making a pleasant day sail from the inner harbor, through the tide rips at Trial Island, around to Oak Bay Marina (the pleasure of our first daytime landing offset by the annoyance of paying for TWO boats at \$.75/foot even though we take up less space and volume than a typical yacht full of people). By the time we launched for the trek to the San Juans, we were feeling attached to the town, and it is a real possibility for the next phase of the project: excellent cruising grounds, active maritime culture, and 2,000

The Maiden Voyage

square feet of proffered ground-floor space in an old shipyard now given over to high-tech companies. Hmmm.

In the meantime, we find ourselves on the northwest end of the islands after another windless 11 mile pedal, in sudden need to move quickly in order to make it back to San Diego in time for a Sony speaking gig, with various essential stops enroute. The thought of rushing through the San Juans is depressing, for we're already getting a sense of the place — island languor mixed with the frenzy of tourism, sharp contrasts between land and sea, potential orcas, and beauty in many forms. But the trip has fulfilled its purpose.

Yes, this has been a significant learning curve. The point of this adventure was to take a reality check — to shake down the boats, camping methods, and even our relationship before committing the next year to Deep Integration of Major Systems. Almost immediately, we started discovering little things — solid tramps are horrible, plastic water bottle cages don't work upside down, boat-to-boat radios need to be built in, an 18-watt solar panel is not enough, Faun needs her own GPS also, and so on. But the bigger realizations grew as we dealt with the daily realities of finding places to sleep and shuffling packs. The Fulmars are exquisite, high-performance little boats for island-hopping, camping trips, and day sails. But for extended travel on the order of years, I don't think they'll do the job. These aren't like bicycles, which allow you to get off for almost all activities besides riding, yet they impose the same packing constraints. My Sony TR81 camcorder, for example, is inside its waterproof case in a Cascade Designs dry bag in the aft hatch... where it has stayed for the whole trip because it's simply too much trouble to raise the solar panel, haul the duffel and lumbar packs onto the tramps, pull the boat bag out of the way, drag the video pack into my lap and unroll the closure, take some footage, then repeat in reverse — especially since so many things worth capturing on video, like that insane first night, involve rough stuff that would make the whole process too risky or troublesome.

Likewise the sleeping issue. Sleeping on board could be accomplished by tenting over a tramp and the center hull with the Kelty Domolite, but that's a fair weather, low wind option only. More to the point is the daily problem of beaching or docking — the former a pain in tidal zones since the boats are too heavy to drag above tide line, the latter twice as expensive for a couple as being in one boat. Again, I've been spoiled by *BEHEMOTH:* I just pedaled to someone's garage or other facility, set security, and slept. We've spent hours trying to moor the boats safely, get our stuff to shore, and relax without fear of pilferage

or other problems. (In Victoria, they spent a night in greasy water near RCMP-impounded bilge-pumping derelict vessels, easily reached by anyone who would choose to dig through them for $500 drysuits and bags of highly marketable marine gizmology. The sludge took eight hours per boat to scrub off.)

A larger boat would let us sleep on board, take advantage of marinas on occasion, improve security over bags and cloth hatch covers, anchor out and kayak to shore, and still beach when appropriate. So now the task is to find one that is still small and light, amenable to the implementation of pedal and solar drive, and possessed of that ineffable magic loosely characterized as high-WQ (Weirdness Quotient). <sigh>

But for now, onward to Seattle... we're taking a day in the Henry Island cabin for me to write and Faun to take a first pass at amortizing $140 worth of fishing gear purchased during a last-minute parking-ticketed frenzy in Seattle (five good salmon dinners-for-two, we figure)... then we're eastbound through the islands and south to close the loop.

Yacht Crash on the Mighty Swinomish

The San Juans were indeed an amazement, though I'm glad we did them off-season. We actually managed to snag one of the four mooring buoys adjacent to the tiny Blind Island State Park across from Orcas, though we had to hitchhike on the dinghy of a neighboring yacht to get ourselves and our camping gear to shore. Hmmm, THAT problem again.

We did do some real sailing, however, a treat after endless exhausting miles of recalling the headwinds and hills of yesteryear. The Fulmars are so low to the water that an eight-knot sail is an adrenaline-pumping thrill, and we scooted through narrow channels, past yachts and ferries, out across open straits, around wooded islands—dancing together and flashing thumbs-up across sun-sparkled water. These are the moments that make the others worthwhile.

At James Island, we camped for a night on a hilltop with a view of Mt. Baker alpenglow to the east and the tiny cove to the west... and the beady eyes of raccoons all around. The first inkling of trouble was the sound of a zipper as we dined on pasta at the picnic table in candlelight—I switched on my Pelican headlamp and sprinted to the tent only to find an entire unopened bag of Fig Newtons gone. These guys play for keeps! A few moments later we heard them scrapping and fighting in the woods...

We didn't know it yet, but the next day was to be our last on the water. After slowly crossing Rosario Strait and working our way down to the famed Swinomish Channel (the alternative to the entirely TOO exciting Deception Pass, whose raging currents sometimes reach eight knots), we enjoyed a tailwind cruise in favorable current all the way to the little town of La Conner. I ducked into the guest dock and managed a smooth docking maneuver, then called Faun on the radio to tell her I had found a spot.

"I'm out of control!" she cried. "My pedal drive is broken!" Drifting with the current, with no steerageway in the tailwind, she bounced off one luxury yacht, crashed hard into another, then clung tenaciously to land and managed to tie up. This naturally triggered a rather bemused reaction from the yachties, whose first and natural response was to invite her aboard for a drink

By nightfall, it had been decided. An inaccessible shear pin on her pedal drive had failed, and it made little sense to attempt an on-water repair. We spent the night on the yacht, borrowed the skipper's car, recovered the truck from Seattle, and made five back-to-back drives to return the boats to their starting point.

MICROSHIP 3.0: THE HOGFISH ERA

And so, convinced that the only solution to the stated practicality issues lay in moving up to a trailerable pocket cruiser, we returned the borrowed Fulmar, put mine up for sale, and embarked on the quest for an affordable multihull that could accommodate my monstrous wish-list of technology while providing enough room for two people to live aboard. By early 1995, we were hot on the trail of our next substrate... once again forgetting the original motives that called for a bicycle-scale, human haulable, sleek little boatlet.

In well-intentioned overreaction, we fulfilled the quest by acquiring and documenting a 30 foot folding trimaran—giving us an eccentric yet fast boat and an unthinkably huge project on our hands. Even though we really had no place to work on it—certainly not the third floor lab at the University—we were convinced that this was the solution. Control and front-end systems awoke on the bench and industry became more involved than ever in the electronics projects, while the TO-DO list on the substrate itself quickly outpaced my available time, skills, and funding...

Heh. That was just the beginning, and before long a new problem emerged—loss of the UCSD lab. We embarked on another epic cross-country combination speaking tour and space quest, and from a pay phone in a Key West bar learned that Apple Computer had come to our rescue by leasing a 2,400 sq. ft. building in the industrial fringes of Silicon Valley. Instead of making occasional pilgrimages to visit *Hogfish* in a dirty overpriced storage yard, she would become the centerpiece of our living and working space! Although it was a windowless, noisy, fluorescent-lit, relationship-stressing concrete tomb under the San Jose airport departure path, it was perfect in a twisted way (and free)... so we hauled Nomadic Research Labs up the coast and moved right in.

By February of 1996, we were into it, making sluggish but steady progress on rigging and fiberglass structures while continuing to refine the on-board network and control tools—video turret, crossbars, data collection, and graphic front end. But something fundamental was missing, and I was loathe to let myself see it.

I did get hints of it during the extended breakup with Faun, still more hints by observing the time I wasted on pointless web surfing, hanging out on CU-SeeMe, and furtively playing computer games. The passion had slipped away, skulking in the night through the grime and traffic of Silicon Valley in its flight to the sea while leaving me saddled with a too-large project that had come to feel like a dead-end job. The volunteer teams that had defined the very heart of *BEHEMOTH* and the UCSD era seemed to have disappeared entirely; the Nomadness mailing list that is my Network community bided its time, saying little, waiting quietly for completion of the yacht or, more likely, the fizzling of the whole mad idea. It had simply gotten too big for one person to manage, and it wasn't really my dream anyway—what, me a yachtie? I was in over my head, working to build something I really didn't want after all and even had trouble imagining.

And that's mighty hard to admit when the world is watching and sending bright-eyed reporters around to do breathless interviews with the high-tech nomad. I simply pushed my misgivings into the background and kept at it for another year.

Flashback to Early 1995

The Microship Finds a New Substrate!

In our last episode, so long ago, we found ourselves aboard a pair of tiny Fulmar-19 pedal/sail trimarans, plunging through the flying spray, soupy fog, unexpected calms, and interesting currents of the Pacific Northwest. Not only was this a memorable adventure, but it substantially changed the very concept of the Microship.

One of the most notable revelations was that full-time cruising for two requires, at minimum, a boat big enough for two to sleep on (and not just in swimming-pool conditions). When you add the need for a robust array of systems, solar power system with electric thruster, communication tools, and enough provisioning to realistically consider passage-making… then toss in decent rigging, essential safety equipment, and a healthy dollop of seaworthiness… you have just reinvented the yacht.

Upon our return from the Fulmar adventure, therefore, we began a quest for the perfect boat (within our budget, alas). I flew to Texas to sail a Stiletto 27 catamaran, talked to sellers of 25-35 foot cats and tris across the land, and collected study plans from noted multihull designers of the Pacific Northwest. We looked at a MacGregor 36 catamaran, sketched interiors, made cardboard models, and found ourselves gravitating toward a sleek Hughes 30-foot cylinder molded tube cat that we could have had built in Oregon… only to discover after drawing a hull cross-section on the wall that it would be like living in a parallel pair of narrow tunnels. We backed off, recognized the structural advantages of a tri, and homed in on a Marples Seaclipper – something that could be realistically homebuilt of plywood and epoxy. But the two-year, $25,000 investment necessary to arrive at the start of electronics packaging was, despite no doubt interesting learning curves enroute, a daunting reality. The calls and research continued… somewhere there had to be a boat that was complete enough to give us a head start but not so complete as to be unaffordable.

Then came a stroke of luck. I was speaking one afternoon with Mike Leneman of Multihull Marine in Marina del Rey, and he was explaining that what I obviously need is an F-31 "kit." For only $52,000, I could buy a sleek folding trimaran shell with no interior or rigging… and hey, it comes with a free trailer! I chortled and told him that I'm a freelance writer who has managed previous technomadic projects through the twin miracles of sponsorship and obtainium, and was about to politely hang up after giving him a capsule summary of *BEHEMOTH* and the *Microship.*

"Wait!" he cried. "I have your boat right here!"

This was every bit as exciting as it sounds, and the net effect is that I am now the owner of an eccentric 30 foot folding trimaran steeped in multihull history… built in 1983-84 by John Walton and

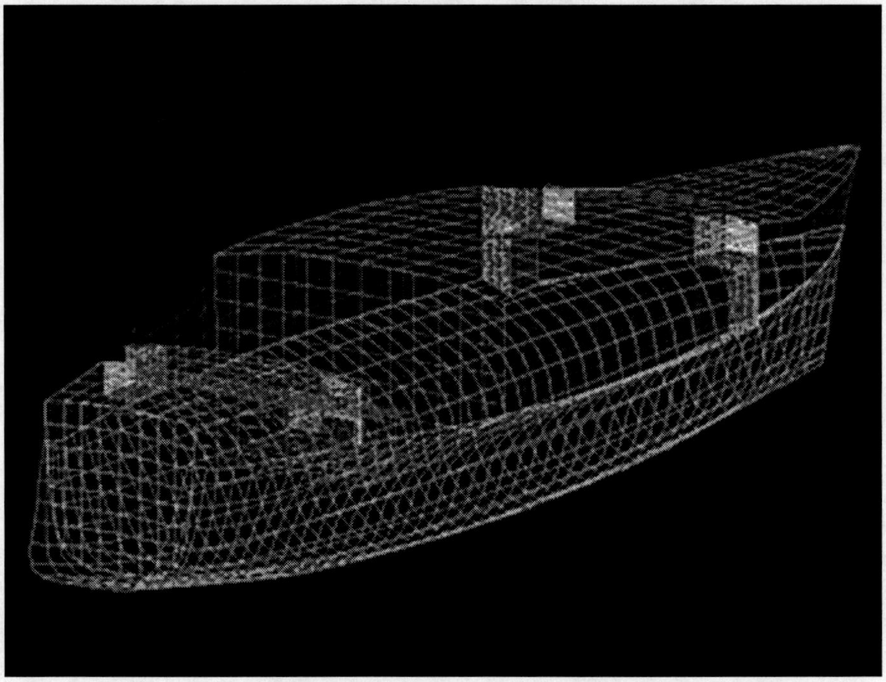

CAD model of Hogfish center hull, derived through a painstaking 'de-lofting' process by David Berkstresser.

Flashback to Early 1995

Mike Michie just prior to the formation of Corsair Marine (builder of F-series trimarans). Affectionately known as *Hogfish*, this 3,000 pound boat has no interior accommodations, lacks rigging, and has languished in dry storage for most of its life—sailed only 4-5 times. The carbon wing mast originally custom-built by Rutan for the boat has become a billboard, but Mike tossed in a used 40´ aluminum mast section and a traveler track to get us started on rigging.

Construction is excellent—S-glass over Klegecell core; carbon-fiber crossbeams, aft cabin, Lewmar hatches, kick-up rudder, and so on. We have the bow and stern pulpits and various other bits and pieces, but are essentially starting from scratch with rigging and outfitting. Suddenly it all seems more real, albeit somewhat intimidating… and we wander by the storage yard every few days to lounge in the cockpit and ponder the enormity of it all.

Hogfish arriving in San Diego CA.

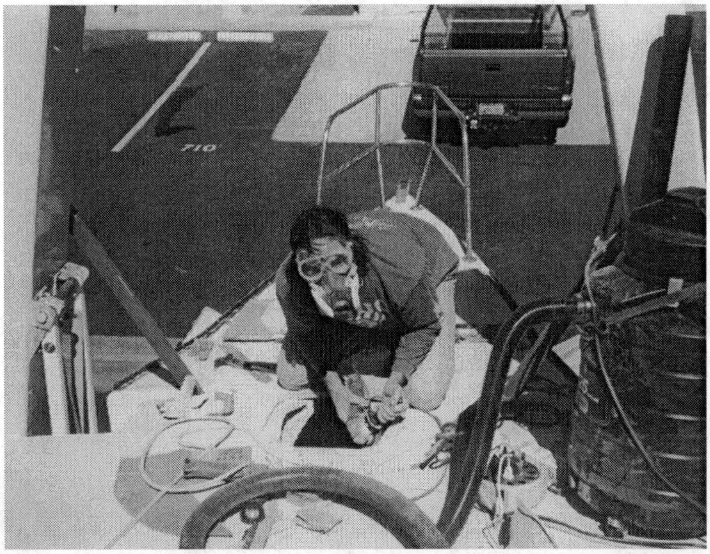

The nightmare of fiberglass hacking…

MICROSHIP 4.0: RETURNING TO OUR ROOTS

But in the Spring of 1997, with Faun gone, Apple funding dried up, expenses skyrocketing, keynote bookings down, and the TO-DO list more intimidating than ever, my world changed completely…

At that instant, the project morphed radically into a mad blend of the Fulmar Interlude and the emerging wireless era. It also happened that on the very same day that I had the epiphany about jettisoning *Hogfish,* an intense email correspondence began with a lovely young woman in London; seven weeks and 1.2 megabytes later, Natasha flew across the Atlantic and we embarked upon a multidimensional relationship that thrives to this day. She commandeered the Draco Casablanca digital video editing system, conjures artwork from some primal taproot of mad imagination that has always eluded my engineering grasp, fears not the epoxy demons, learned PageMaker and carried the first printing of this publication from raw text to final layout in under a week, and can whip up a mean curry. She passed her amateur radio technician test (KF6NWO, a callsign rich with apropos acronyms), and at the moment is assembling and interfacing a Paia theremin and FORTH microcontroller to experiment with gestural control of the video suite being built into her Microship. Natasha is the Nomadic Research Labs miniskirted geek-babe, and as the resident Old Fart, I still can't quite believe my luck.

We sold the big boat and began work on *Io* and *Europa,* a pair of tiny canoe scale micro-trimarans that integrated all our existing electronic systems while trimming two tons of nautical fat. There was collective rejoicing from my Internet mailing list, now about 2,500 strong. And in early 1998, with *Io* almost ready for test sails, we escaped Silicon Valley and relocated the project to our own 3,000 square-foot lab nestled in a 11-acre wooded home base on Camano Island, WA… complete with a passive-solar house and nearby launch ramp. And it is here that we are pushing hard toward completion so we can actually travel—soon!—in these wondrous machines.

And so, this historical chautauqua has brought us at last to the present, a full-circle of sorts, but with so many learning curves enroute that it all seems necessary in retrospect. We are fortunate to have a guru-level fiberglass fabrication expert on the team, and now that the scale has returned to insanely weird ultra high-tech amphibian solar-encrusted micro-boats bristling with antennae and aglow with blinkies, volunteers have again emerged from the vapors of the Net to apply their wizardry to everything from power management to linux server hackage.

But before delving into the details, let's indulge ourselves with a flashback to the moment when the Microship project veered back on course…

SKR with Natasha at Sea Kayak Symposium, Port Townsend WA.

Flashback to Early 1997

Silicon Valley, CA

Embarking on a Non-sequitour

It was Sunday, March 23, and I paced the lab like a caged animal, constrained to a windowless rectilinear space with artificial light, vague dreams of aquatic adventure burbling in my reptilian brain like something instinctual driving me to odd behavioral sublimations that give vent to long-forgotten urges. At the center of the lab stood the boat, Icon of the Dream, covered with dust, tools, respirators, Shop-Vac, and the funky detritus of fiberglass hackage. Surrounding the boat on all sides were cluttered benches, shelves laden with expensive nautical and electronic hardware, uncategorizable clutter, chip databooks, sticky epoxy fixin's, a maze of cables, computers, and stuff. I paced. I climbed the steps to the cockpit, paused, and descended. I idly picked up the AIR Marine wind generator and spun the shaft to make the LED flicker. I peered through a port into the aft cabin and stared at the battery bank, putzed ineffectually with the flute for a moment, and scanned the "Fiberglass Projects" list densely occupying one of four whiteboards, hoping as always that I could erase something… but no. I randomly picked up last April's issue of *Sail* magazine and thumbed the pages, absently, as one might flick one's eyes over a rack of postcards at a checkout counter.

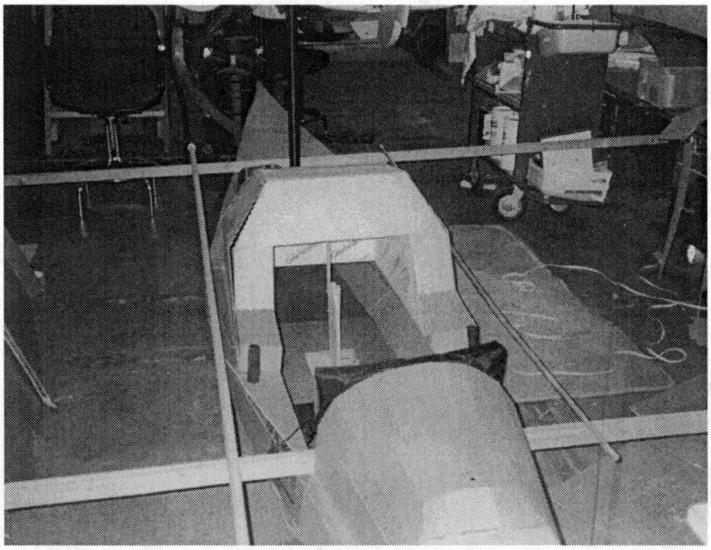

The CAD System (Cardboard-Aided Design).

Understand, we're talking major restlessness here. A half-decade in Microship labs or on a quest for same; the bike (the world's most expensive boom box) playing jazz while mercilessly taunting me with road-dirty wheels and a design optimized for movement. Ah, memories <creak>. Feeling like an old timer lapsing into reverie, I ignored the relentless itch of instincts unfulfilled, consciously turned my back on the TO-DO list, slumped on a lab stool, and started browsing the magazine.

Ah yes, that article. One of my favorites – the comparison of "sailing kayaks": the familiar Fulmar 19, the fast rotomolded WindRider, the Balogh Batwing, the racy foiled Triak, and the pretty wood Chesapeake Light Craft SailRig… all sleek, light, fast, exciting little boats. <sigh> Aching images of the Fulmar adventure with Faun bubbled past my barnacled brain as I read the piece again, lingering on the pictures, recalling the fantasy that started all this back in 1991… pedaling *BEHEMOTH* up the Wisconsin shore of Lake Michigan while fantasizing about gizmological boatlets… but no, no, stop thinking about that! I've been there; the Fulmar was too small, THIS is the boat, this is the boat I have to finish NOW. I looked, almost guiltily, at *Hogfish* and felt… nothing.

Wow. Never really noticed before, but gee, now that I thought about it… I really haven't had much of an emotional connection with this boat, have I? It (not "she") is a good logical solution to the stated design goal, which had something to do with a trailerable platform for a happy couple on the technomadic adventure trail. But what about the passion, the heart of this whole project? Suddenly alert and suspicious at the pheromonal scent of Major Change in the air, I stared at the boat, then back at the magazine… then back at the boat. The toggling between adrenaline and yawns was too damned obvious to ignore, and I was instantly embarrassed that I hadn't noticed this a year earlier.

Reality Checkmate

When this whole Microship project began, you see, I perceived it quite simply as the aquatic equivalent to my computerized bicycle. Human/solar, way geeky, challenging, light, physically minimalist, small enough to haul out and stash in a host's garage, and (unlike yachting) approachable by people in the sense that they could imagine themselves doing it. I drew countless sketches of micro cats and tris, pedal powered, with a road mode based on pedaling or walking. Then, as the wish list *(The BEHEMOTH Effect)* began to kick in, the boat grew… and grew… from a kayak to a kayakamaran to a kayacht… to a 30-foot center hull with double kayak amas… to a brief fantasy about a ridiculous amphibian diesel motorhome with hydraulic crossbeams… and then, almost as a backlash, to the Fulmar… which was wonderful but too small for the growing complexity of a project that was not downsized to match (couldn't haul out or sleep on-water anyway). I had a brief adventure and sold it, embarked on a quest for the ideal trailerable live-aboard-for-two folding multihull, and settled at last upon this unique and elegant precursor to the sleek Corsair F-boats: the one-of-a-kind *Hogfish.*

Two years passed.

During that time, I continued building my electronic and control systems, which were to occupy a 10 foot long, 6-bay pressurized console in the cabin. I acquired most of a 40′ rotating rig and puttered at boatbuilding, completing the new bulkheads and mast step assembly as well as a few minor additions… but recently the ratio of untouched projects to lab time remaining reached a critical level, and I had to accept the fact that at least a year of work and $20,000 yet separated me from those first tentative non-electronic test sails. Only then could I begin integrating the comm/nav and computer systems that underlie the whole technomadic dream—the passion that drove me to do all this in the first place.

But when I wandered the US, much too long ago, it was by bicycle instead of motorhome: the comfort and insulation of an RV would have been stifling; the culture, however pleasant, boring; the mountains and deserts, too easy. A direct comparison to sailing is impossible since there are additional issues of safety, wind, sea state, and the small matter of being unable to step off at one's convenience to take a break… yet there is one relationship that I have experienced enough to believe correct: the smaller the boat, the bigger the adventure. Since smaller boats are also cheaper, quicker to build, and (if carried to the logical extreme) light enough to schlep around under human power, we can postulate another solution set to the complex formula that is the Microship.

Before you throw up your hands in exasperation and imagine another five years of reading Nomadness tales about geek adventures at the helm of a fluorescent-

FLASHBACK TO EARLY 1997

powered workbench, there's something I have to tell you: All the electronics work already done transfers directly, while the number of projects remaining drops by about 75%...

Now that is tempting!

Canoe's Not Unix

Okay, now fast-forward six weeks. The lab has a whole different feel, more incubator than cage; instead of pacing, I peck vigorously at my shell while working on a much cheaper project. *Hogfish* is being gussied up for the trip to a multihull broker next week (unless you want to buy a low-cost folding trimaran and contact me immediately, that is!), and the dominant feature of the open workspace is now an actual-size CAD (Cardboard-Aided Design) mockup of one of the new boats.

The new Microships are tiny trimarans based on Wenonah canoes, being built in Minnesota with a custom layup to accommodate some decidedly non-canoe-like stresses. These are lovely hulls, and will be decked over with foam-core fiberglass and bulkheaded to add strength and seal off the bow and stern storage compartments. The outer hulls and akas are stock units from Fulmar Canada, with the entire spaces between them covered with folding solar arrays. We unpacked 16 of the Solarex 30-watt modules and laid them out on the model... and hey, this is pretty!

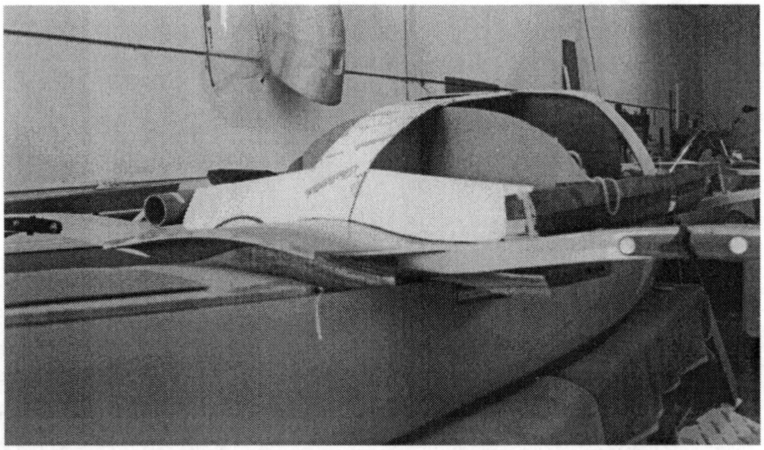

First stage of folding console design.

Having two boats that weigh perhaps 600 pounds each instead of one that weighs 3 tons makes a huge logistical difference: with pivoting fat-tire wheels supported at the bulkheads, it is a simple matter to fold up the akas and turn the Microships into manageable trailers on a moment's notice. With one decision, we have eliminated dependency upon the whole nautical infra$tructure of marinas, ramps, boatyards, and a dedicated truck/trailer combo.

The operating position is reminiscent of *BEHEMOTH*—recumbent seat, pedals a bit higher, sliding armrests linked to the rudder by hydraulic cylinders and supporting computer-interface hardware, and a wrap-around pressurized equipment console that houses all the computer, nav, and comm gear. A full bimini cover (like a convertible top) can be deployed over the entire cockpit region, supported at the rear by an arch and forward by a Lexan window atop the console.

One of the most pleasant and unexpected aspects of this downsizing is that it caps the proliferation of electronic systems at a reasonable level... it actually took me a few weeks to accept the fact that I was already grasping the entire power distribution and control network. Nahhh, it can't be THAT easy... but suddenly it is. A lot of redundancy has gone away, and even such easily overlooked but critical issues as single-point grounding have gotten a lot simpler. I realized the other day with a bit of a startle reflex that the network running on the bench is a very large percentage of the whole system, rather than just a tentative start.

Finally, I have to say that this simply feels right. I've delayed writing this report for a lot of reasons... this amazing mailing list includes most of my friends, sponsors, critics, volunteers, consultants, and distant acquaintances; posting something here is always a big decision. This project has gone through a lot of changes since my first tentative maunderings about computerizing a kayak back in 1992, and reactions have covered a wide range: encouragement, impatience, disagreement with technical or lifestyle choices, technical suggestions, vicarious amusement, and quiet tolerance of a long project. I look back at the early tales, my impassioned discoveries of magic waters intertwined with unrealistic but well-intentioned notions about boatbuilding... and over the half-decade since I have watched my education grow while adventures languish. But now the curves have crossed; I know enough to finish this simply and get back to where I belong—even if it does involve one last change of substrate.

And THAT, my friends, is what this is all about. Five years in the lab has turned me into exactly what I fled so long ago, while out there, not really so far away, are quiet coves sparkling in the moonlight, mist-shrouded rivers sussurating through canyons of naked rock and verdant green, sunsplashed beaches that beckon like eternal visions of idyllic summer vacation, campsites abuzz with chittering and aglow with firelight, those first long swallows of cold beer after an all-day haul, the spark of delight in the eyes of a suntanned friend, the rush of a reach that hardens the sail and foams the wake, the magic correlation between dead reckoning and a fix, the astonishment of a satellite email packet, the crackle of a DX contact in the wilderness, the steady blinking of LEDs that bespeak volumes about the systems that drive them, the tales that fly when nomadic spirits recognize each other and pause to share the calm moment of sunset, the adrenaline-pumping madness of nature's fury, the love, the joy, the perspective, the LIVING...

Maybe big boats do make more practical sense, but I don't care. It's time; the ancient urges are surfacing in a sort of Spring Fever of the Soul. Life is passing; pheromones are in the air; there's restlessness in the night.

Everything else is secondary.

Microship Design Goals

Obviously, I've had quite a while to think about this. On the surface, the Microship project is an outgrowth of the technomadic bicycle adventures that somehow became a career: the cyclic creation and application of tools that allow me to remain productive and connected while traveling freely. With physical location reduced to irrelevancy, it follows that I can roam the planet without regard for deadlines, dwindling bank accounts, or schedules beyond those imposed by the seasons.

There are some non-trivial components to this, many of which were not at all clear when I started wandering on the bike. Over time, the specifications evolved, shaped by reality and changing needs in just about every domain: connectivity, computing power, power generation, maintainability, design flexibility, physical security, multimedia tools, scaling gracefully to accommodate a flotilla of other travelers, and more...

After reviewing the complex history of the project and seeing how easy it is to get blown off course, we should now clarify our thinking by stating, as concisely as possible, the point of all this:

The Microships are intended to serve as physically minimalist escape pods for open-ended adventure along coastal, inland, and protected waterways (not ocean crossings or whitewater). Sleeping on-board must be possible, albeit perhaps spartan, and it is essential that haulout can be accomplished under human power without the need for trailers or other shore-based facilities.

So far, this spec could be met easily by off-the-shelf kayaks or canoes... but there are a few other issues to consider that introduce compounding complications such as the need for deployable landing gear:

We require multiple independent modes of propulsion in the form of pedal drives, steerable solar/electric thrusters, and freestanding sails. A sealed, pressurized control console is needed on each boat to accommodate on-board computers, communication, and data collection tools. One of the boats needs full scale digital video editing capability, as well as a suite of on-board cameras including a steerable video turret. Satellite and HF data communication facilities are essential to maintain our links to the Internet for email and telemetry, not to mention a capable ham radio station, cellular phone, marine VHF, and so on. A full set of navigation tools is required, along with "smart backpacks" to allow peripatetic crew to locate each other or the boats... and a wireless network must be continuously available to allow full graphic remote control and monitoring of Io and Europa even when they're on the beach and we're in town. And, of course, the whole system must be exquisitely hackable, exciting, robust, and capable of sustaining not only basic life support but high levels of amusement for a long time to come.

ADVENTURE PLANS

We can further refine a sense of scale and direction by moving from the rather abstract specification above to a statement of where, exactly, we plan to take these machines. By mapping the two disparate sets of constraints upon each other, something much like a design should begin to emerge.

One of the primary reasons we located the final staging area of the Microship project on Camano Island is ready access to some of the most spectacular waters in North America… Puget Sound, the San Juans, Gulf Islands, Princess Louisa Inlet, and more. The entire Pacific Northwest is a favorite cruising destination, yet lends itself to small-scale gunkholing quite unlike San Francisco Bay or the open coast of California. Our test sails and feasibility studies, therefore, are all taking place here.

Bike experience has proven that a shakedown cruise of significant scale is essential before taking off on a grand adventure—the need to rely completely upon on-board systems is never adequately tested by stringing together local day trips. Given our location, the obvious choice is an epic loop of the entire Puget Sound area, and well into Canada. The current plan is a solo early-summer "Previews of Coming Attractions" (PCA) tour, encompassing a variety of on-water events and interesting navigational challenges. Shortly thereafter, a cross-country speaking tour via our new Mothership should provide additional opportunities for mini-adventure in a variety of aquatic settings, giving us a final reality check before we dive into the final six-month system integration and packaging project.

With all this at last behind us and the lab mothballed, we'll be off on our first serious adventure… a circumnavigation of the eastern US beginning just down the road at the mouth of the Columbia River, at Ilwaco, Washington. From there, it's a 465-mile slog upriver to Lewiston, Idaho, where we'll load the boatlets into a truck and haul them over the Rockies to the headwaters of the Missouri River at Three Forks, Montana. (A friend pointed out recently that since we'll be tracing the Lewis & Clark route in reverse, we should call this the Clewless & Lark Expedition…)

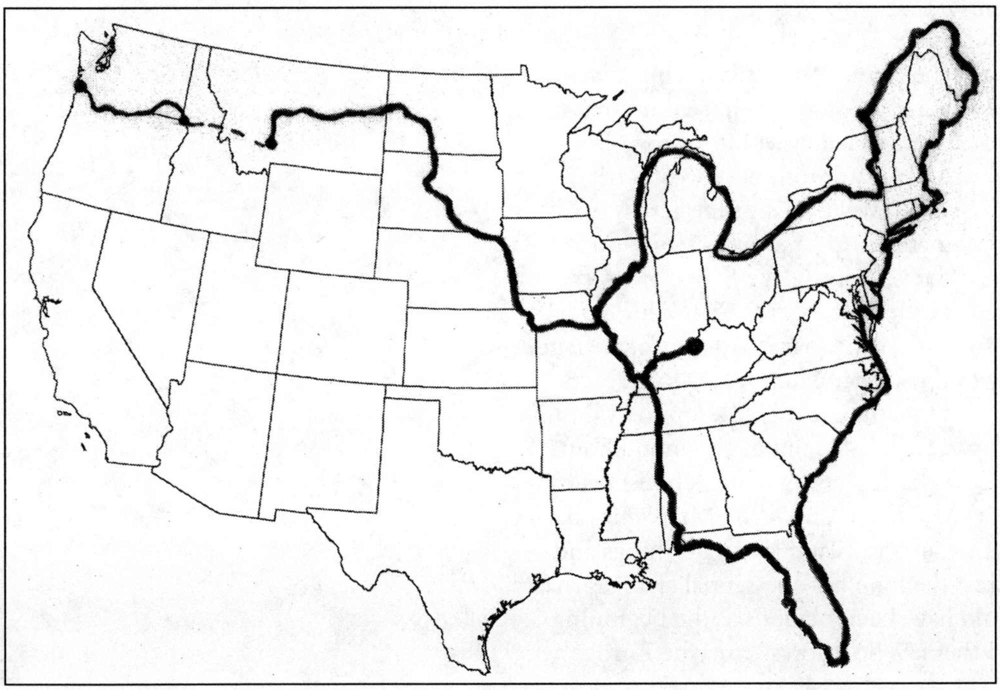

Planned Microship expedition around coastal and inland waterways of North America.

Adventure Plans

Anyway, there begins the downhill run, though from what I've read of the upper Missouri that's a bit of an oversimplification. We'll traipse 2,546 miles across the northern plains and down between Nebraska and Iowa… cutting east to join the Mississippi River just above St. Louis. A short float downriver brings us to the mouth of the Ohio… then upriver slightly to the Tennessee, whereupon we turn south and continue along the Tenn-Tom Waterway and down the Tombigbee River to Mobile, Alabama (the more obvious parallel path down the lower Mississippi is relatively hostile to small boats). At the Gulf, we turn left on the Intracoastal Waterway and meander all the way around Florida (through the Everglades and the Keys) and north along the Atlantic Coast. The ICW will carry us past Boston with no shortage of interesting exploration enroute.

At this point, unless forced by seasons or sanity to shortcut up the Hudson, we encounter open coast for a while as we traverse the exquisitely convoluted Maine shoreline—then into the Bay of Fundy followed by a short portage across New Brunswick at Moncton (to skip the suicidal outside coast of Nova Scotia). After darting around the Gaspé Peninsula during a favorable weather window, we'll head up the St. Lawrence, turn left at Montreal to sail down Lake Champlain, pop over to the Erie Canal, travel back in time across New York, emerge into Lake Erie, cruise up Lake Huron, wander into Lake Superior unless we elect instead to cut through Chicago to the Illinois, head upriver at Duluth, portage over to the Upper Mississippi, float down Old Muddy, then finally slog up the Ohio River to Louisville, my old boyhood home, where we'll stop at my father's house and truck the tattered and filthy Microships back to our Camano Island lab to fix things and address the long list of essential changes that should have been obvious at the beginning. And then? Who knows… maybe Europe?

It's clear that a wide range of conditions will be encountered—infestations of mosquitoes and jet skis, high tidal currents, storms, urban waterways, exposed coasts, humans of uncertain motives, and things we don't even know about. Every day on the water is different. Throughout it all, we need to be able to accomplish daily haulouts as necessary by deploying landing gear, camp on-water when shore options are limited, resupply every few days, switch propulsion modes as conditions dictate, and manage daily life support… all while presenting the illusion of stability to the world by utilizing our technomadic systems to publish, consult, interface with the educational community, collect and post environmental data, and run the strange business that is Nomadic Research Labs.

A Brief Rhapsody on Art and Engineering

It is essential, when designing a complex system, to spend some relaxed time fantasizing about what it will be like when it's finished. After all, this is what drives the process of engineering: at some level between rigorous and fanciful, an image of the finished product must be held in the mind, savored, and examined from all sides. Only after this playful interlude (which, to a manager, may be disturbingly indistinguishable from unproductive wall staring) can decomposition of the design into subsystems, tasks, and packaging make any sense.

Trying to shortcut this by starting on Day One with formal design methodologies can have the catastrophic effect of committing one to an ill-defined goal state, whereupon the end result is shaped more by design tools than the supposed objective. That's why so many products seem malformed, patched, and otherwise inelegant… the industry loves formal tools, and generally looks askance upon such frivolous notions as approaching product design as a delicate blend of art and engineering. The exceptions, when they occur, are a joy to use. The rest simply miss the point, no matter how stylish their exterior… or how sophisticated the underlying technology.

With that in mind, let's play with Microship system design at the highest level for a moment. What will this feel like? Perhaps a "day in the life" approach, though a bit clichéd in a literary sense, will help integrate all these processors, networking layers, and distributed widgetry into some kind of a cohesive (and hopefully enchanting) whole.

It works! First test sail, Port Townsend WA (October 1998).

A Day in the Life of a Technomad

My dreams are scattered by an Aaron Copeland trumpet fanfare that's so absurdly grandiose in the context of a wilderness campsite in British Columbia that my partner and I simultaneously giggle awake and grin at each other. The music fades before the orchestra has a chance, then I hear a subtle click as the audio crossbar pipes the speech synthesizer to the Microship speakers.

"Good morning, campers... time to get up! Local time is 7:15, ambient light levels suggest sunlight, and the temperature is 58 degrees. Eight email messages have accumulated overnight, the main battery is at 91%, your portfolio is up 4% in morning trading, and all internal ship systems are within normal operational limits." Somewhere in the heart of the Microship, an MP3 file starts streaming from disk, gets decompressed on the fly, and manifests itself as a delicious, spacey piece from Air entitled *All I Need.* I smile, seeing Pixel Toy graphics in my head, recalling nights in Amsterdam...

We wake and stretch, cozy in the tent nestled between *Io* and *Europa*... a pair of amphibian micro-trimarans parked parallel in the grass. Above us, a graceful custom tarp supported by retracted amas serves as rainfly and sunshade, and when we zip open the tent door it's like looking out through a giant portico. As every morning, it takes a moment to believe where we are: gazing out at Gabriola Passage in the Gulf Islands, at last living the technomadic life that's been the driving passion behind eight exhausting years of development work.

While espresso hisses on the fold-down propane stove and Java the cat makes her morning rounds, we lazily do the prefloat. I fire up a browser on the PowerBook and connect via wireless ethernet to the Microship on-board web server. The morning report was about right—no emergencies today (not like last week, when I woke to discover that console pressure was zero and the pneumatic controller had crashed... a potentially catastrophic situation in a salt-water environment). I quickly review strip charts of the power management system, scan the log of overnight telemetry transmissions to the homebase server, and open Eudora to breeze through my email—fetching it from the IMAP server on the boat.

I open a new outgoing message: "Heading south today for Cowichan Bay. Just breaking camp in Drumberg Provincial Park. All systems normal. More after we launch... I'm still uncaffeinated." This will go out during the next automatic hourly connection via Globalstar satellite, accompanied by an internally generated telemetry block that will add the latest snapshot of 50 or so time- and location stamped sensor readings to the browsable database on our public website, along with a bucolic still image from the video turret on Natasha's boat (currently pointed across the Passage).

Over latte and breakfast, my sweetie and I discuss the day's run, which will be a bit critical in spots due to tidal currents. We've learned the lesson a few times in these waters... fighting the current is pointless and frustrating; working with it is like cycling with a tailwind. I don the wearable and bring up the chart and tidal data, emailing us both a few notes about waypoints and timing targets that can be popped up later on our on-board browsers.

But now—there's no time to linger! We pack up the camping gear, kiss for good luck, and roll the boats down the beach on their deployable landing gear. I always like this part… smooth transition from land to water and back again was one of the key design goals, though for quite a while we had our doubts (going so far as to publish a story entitled "Why Boats Don't Have Wheels"). We still have to struggle when we encounter goopy mud and rocky beaches, but this isn't too bad… in a few moments the boats are bobbing in gentle lapping wavelets and we're retracting the landing gear. A few more quick checks and we're off.

There's not much wind at the moment, so I keep the sail furled, deploy the SpinFin, and pedal into deeper water.

Here's where it all kicks in, and there hasn't been a launch yet that fails to deliver this feeling. I slip on the lightweight Xybernaut headset and bring up a zoomed-in nautical chart on the heads-up display; it's oriented to my heading and shows my own location as a little boat icon, slowly tracking my movement. The console browser wakes up and displays the main monitor screen, with live power, network, diagnostics, comm, and sensor readings. All objects have a click-through that displays more detail—I crawl into the power management system and look at current measurements around the boat, check the power budget parameters, and see that we're getting a healthy 16% gain from the set of eight peak power trackers. I note with satisfaction that the cool weather and bright sunlight are already giving us an estimated power surplus of about 210 watts, enough to use the solar drive.

I turn around and deploy the Minn-Kota thruster, and set the console throttle to maximum free power: until I request otherwise, it will divert as much energy to propulsion as it can without impacting battery charging or internal system loads. I retract the pedal drive and kick back… watching the world silently slide by with only the subtlest humming and sussuration

of water on the hull revealing that we're cruising at four knots on sunlight alone.

Natasha's voice purrs from the stereo speakers in the arch behind my head. "This is beautiful, Steve. Check out the video—I'm sending it over on the analog channel." I click on a tool that opens the video crossbar chooser, and pipe the incoming material from *Europa* to the auxiliary console display. Nice shot indeed—she has the turret gazing back at herself, and I see a pretty, grinning woman peeking at me through a polycarbonate windshield that's reflecting the low mountains behind Nanaimo… all surrounded by bright rippled sea foamed lightly by her wake. I send back a live grin from my console camera, and a few hundred feet away on her matching trimaran she snags snippets of both channels onto digital tape for later copying to the Casablanca—her onboard digital video editing system. With eight cameras between the two boats, including underwater units complete with hydrophones, there's never a shortage of source material.

I pull a waterproof keyboard onto my lap (preferring it for speed and comfort over the chording unit embedded in the hydraulic steering levers), and work on the day's email, then do a quick edit on a long overdue article for *Dr. Dobbs Journal.* I seem to be more productive out here, for the noise level of deadline driven life in the lab is but a memory and the writing process is intrinsically amusing. "I'm in a canoe," I giggle

A Day in the Life of a Technomad

to myself, tapping away as the islands drift quietly by. Suddenly thirsty, I pluck my insulated mug from its nest under the gunwale, hold it under the tap, and twist a knob: a little Shurflo pump whirrs to life and delivers a cold drink from a seven-liter tank. Ahhh, such decadence…

The morning passes lazily, punctuated by a bit of ham radio chat with folks on Vancouver Island and a discussion via Globalstar with one of our technology transfer partners about a spinoff project. We decide to stop in Chemainus for lunch, so we tie up at the public wharf, set security, grab the manpacks, and stroll into the pretty little town… enjoying the stretch. Sitting in a sidewalk cafe, I'm just starting to relax when my pack beeps insistently and the dual-band ham radio crisply utters, "Microship security alert: starboard microwave proximity sensor triggered at 1417 hours."

I don the wearable headset and key it out of sleep mode… open Netscape, and click the HOME button to connect to the website aboard *Io*. It takes only a moment from the welcome screen to access a webcam tool that lets me switch among the available cameras. Natasha does the same via her laptop as I look around, scan the area with the steerable camera on the forward winglet, and—aha! A couple of teenagers are enchanted with the solar panels and one is sitting on the dock, nudging them with his foot. Sheesh… no doubt harmless, but still… I click a button to power up the video recorder and start recording this channel, then turn on the laser adjacent to the camera, fine tune the Az-El controls to place an intimidating spot on the offender's chest, open a channel from synthesizer to stereo, select a firm male voice via a pull-down menu, and crank up the volume. Finally, I type, "Remove your feet from my starboard solar array or I will begin reciting poetry!"

As the machine utters its ridiculous warning, he almost falls over himself in his haste to get away, looking around guiltily as other boatowners peer over to see what's going on. The last we see of the pair, they're sauntering off with feigned insouciance, hands in their pockets, followed by both my Az-El camera and Natasha's video turret.

She wants to check out the town and I feel like being lazy, so we part company after lunch and stroll our separate ways. I've almost dozed off on the beach when I realize we're pushing our luck with the currents, so I grab the handheld rig, key the mike, and intone, "Computer. Where's Natasha?" Back on *Io,* a task has been performing great circle calculations on GPS location beacons transmitted every 30 seconds via packet radio from both of our packs, and after my request is processed by the speech recognition system a soft female voice responds, "Natasha is .34 miles away on a bearing of 312 degrees." I give her a call and remind her of the time, then return to the boat to get ready for the day's final leg. On the nav screen, I can see her progress overlaid upon a detailed street map… pausing once as she's drawn into an interesting shop and loses GPS coverage.

As we hoist our sails to catch the afternoon breeze and ride the current down Stuart Channel (congratulating ourselves on good timing—it's not always this easy), I amuse myself with the telemetry system, watching tiny variations in water quality measurements and trying to correlate them with the surrounding environment. This is a lost cause in fast moving waters, of course, but it helps the time pass… I get excited by a dip in dissolved oxygen and spike in turbidity as we pass a smoking mill and concoct a few theories, but it's going to take more data than this. I flag it for later study and add a note to the file that will go out in the next hour.

As Seth Ceteras recently observed via email, I have become a sort of roving probe. "I imagine your boat floating serenely down a lazy river," he wrote, "and in addition to the usual physical wake in the water, there is a 'data wake,' a web accessible data stream flowing back through all the time spaces the boat had just, and indeed ever, inhabited."

Ahhh, technology. We post our telemetry to the GIS-based server at datawake.net and are urging everyone else with live environmental feeds to do likewise… in time, with input from thousands of boats, bikes, trucks, planes, and people, this might yield a rich fabric with deep history that reflects the evolving health of the planet.

We end the day in Cowichan, cruising up the bay at sunset and arriving at the town dock to find net friends waiting to spirit us off to dinner. We had given them a shout via terrestrial cellphone a few miles out, so the timing is perfect. After a brief demo, we stroll up the dock to town… but I pause and look back at *Io* and *Europa*—graceful little canoe-based trimarans resting silently on the water in the day's dying light. Packed with computing power and linked via satellite to the Internet, they're not at all what they seem…

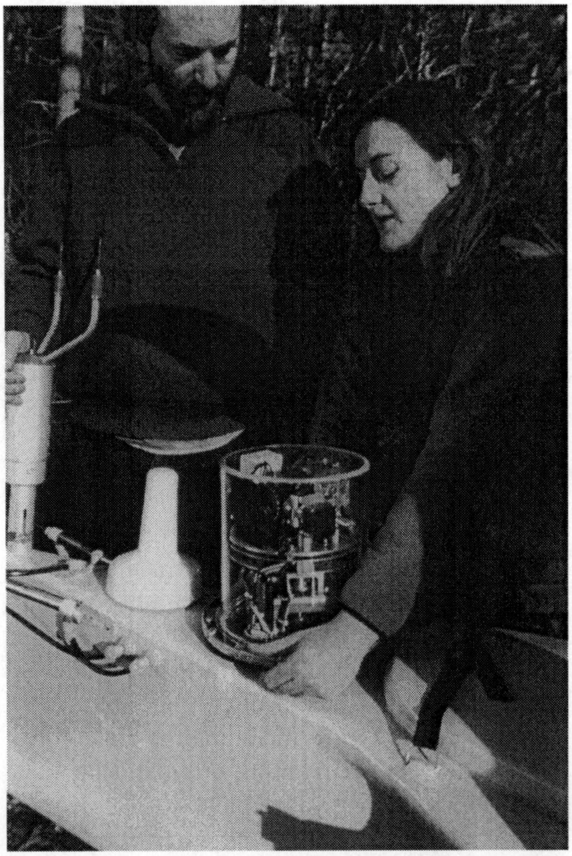

Pondering possible location of the video turret.

Microship Physical Substrate

Okay. Let's turn all that into reality. The first issue, and one that provided many years of entertaining distraction, is just what, exactly, is going to float this whole affair. The substrate has taken on quite a variety of shapes and sizes during the comparatively linear development of on-board systems, culminating at last in the twin micro-trimarans now under construction in our island lab.

The issue of scale is a critical one. It may seem most rational for a traveling couple to occupy the same boat, and as I concluded at the close of the Fulmar epoch there are indeed some compelling arguments for this: companionship, the ability to rotate watches and share difficult tasks, avoidance of equipment redundancy, logistical simplification at marinas, and so on. But the Microship project specification also includes the requirement for unassisted haulout under human power at any time, including the possibility of limited overland transport... and any boat big enough for two people to live aboard full time (without killing each other from over-crowding) is simply too large to be moved without a trailer and tow vehicle. In addition, the whole "aesthetic" that renders this adventure philosophically congruent with the computerized recumbent bicycle that preceded it calls for small personal machines powered by pedals, solar panels, and sail.

Early in the fabrication process, we depended on cardboard models for visualization.

Given all that, our key nautical issue became one of finding (hopefully) or building a pair of small trimarans. One thought was to revisit and hack the Fulmar-19, but as I discovered during the 1994 adventure, the inability to sleep on board or haul a loaded boat above tideline without two or three additional strong bodies led to all sorts of daily headaches... and repackaging the boat to solve those problems would be daunting. Other commercial trimarans of the same general scale, while endlessly amusing to daysail, share the same basic problems.

So... we had no choice but to build our own, hoping to violate the aphorism that the average completion time of a homebuilt boat is 130 years.

Fortunately, we've had the opportunity to steal from the best. The Microship center hulls are based on Wenonah *Odyssey* canoes, fashioned of Kevlar using their ultra light construction method with foam core bilges and stiffening ribs. Our own custom decks, laid up with ten ounce fiberglass and West System epoxy over ½-inch Divinycell PVC foam core, integrate the canoes into the complex system of stresses and fixtures necessary to accomplish our goals; added plywood bulkheads at the high stress areas stiffen the entire structure while giving us sealed forward and aft stowage bays accessed through gasketed hatch covers.

Crossbeams (A.K.A. "akas") and outer hulls ("amas") are off-the-shelf units from Fulmar Canada—life is too short to reinvent the wheel! The scale turned out to be ideal, so we simply boxed in the U-section akas temporarily with scrap Formica, pulled a mold, and integrated the matching nests into the decks. Likewise, we were able to shortcut the whole rigging can o' worms by applying stock WindRider units—elegantly shaped medium roach sails on freestanding aluminum masts, which we stepped in Teflon anodized aluminum tubing glassed to the forward structural bulkhead. To simplify furling, we added a drum for remote mast rotation from the cockpit; the boom, mounted to a gooseneck bearing, is sheeted to a Ronstan traveler on the "arch" behind the seat.

The remaining basic sailboat components are the appendages—and again we worked around available components with the guidance of multihull marine architects John Marples and Jim Antrim. The rudders are carbon fiber foam core units made for us by Moore Sailboats, hung from integrated pintles via a custom hydraulic kick up assembly and controlled via two more pairs of Clippard cylinders by a pair of pivoting aluminum T handles… which also carry computer interface hardware and a few other controls. The boards aren't quite so sexy—they're clunky used Nacra daggers picked up from a sailing shop in Berkeley (hopefully to be cloned in the spirit of the ultralight rudders after initial testing). These are perhaps the weirdest looking part, from a boat design perspective: in order to support the fundamental requirement of sleeping on board, they are deployed through molded trunks at the turn of the port bilge… angled forward in a completely non-intuitive way to maximize the chance of favorable retraction on grounding and to simplify manual retraction from the cockpit.

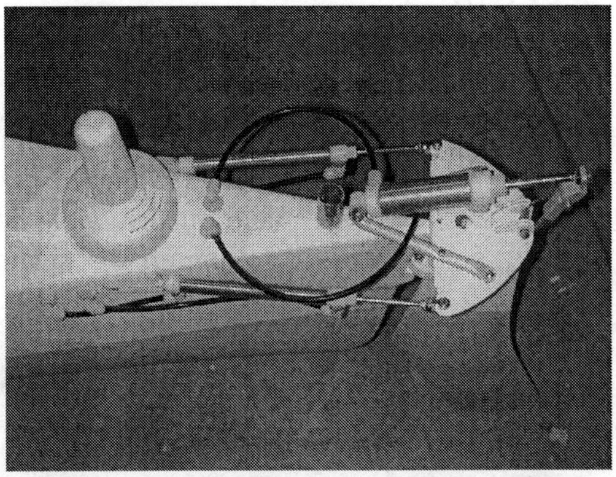

Rudder hydraulics: Clippard cylinders back-to-back with actuators in the cockpit.

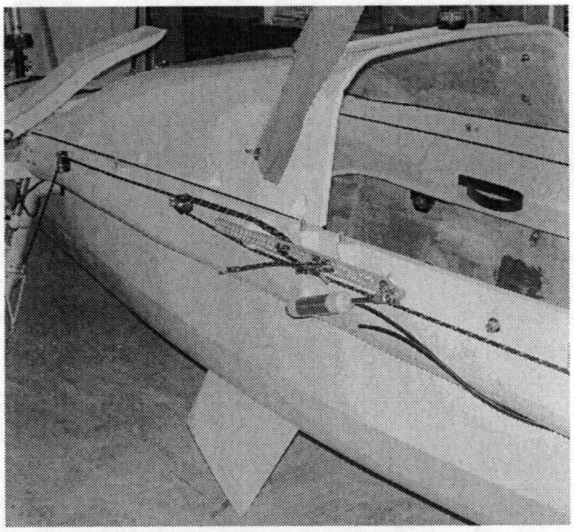

Tech note on the daggerboard: Positioning the board to provide a bit of weather helm is critical, for the opposite can get you into trouble in heavy airs by falling off into a high-stress condition that can cause capsize or dismasting if you don't instantly uncleat the main. Heading up when rudder control is lost depowers the boat and leaves you safely – albeit embarrassingly – in irons. Accomplishing this requires the below-water CLR (Center of Lateral Resistance) to be slightly forward of the above water CE (Center of Effort). As each is the complex resultant of many factors that can themselves vary with wind conditions and reefing, I never felt too confident in our analytically derived daggerboard placement… which is why we used a temporary rudder lashup for the first on-water test (leaving ourselves one last adjustment before having to tweak the helm balance with a Sawzall) But all worked as planned, so we went ahead and finalized the rudder mounting.

Beyond the aforementioned sailboat components, all that's necessary to get the boat on water is a place to sit, a deckload of rigging components shuttling braided line to and fro, and a touch of wind. But many more design issues contributed to the shape of the physical substrate…

First, the on-water bivouac requirement dictates bulkheads about eight feet apart, conveniently matching the spacing of our crossbeams. The adjustable seat, based on a webbed recumbent bicycle unit made by Ryan, can be flipped back onto the afterdeck to allow a camping mattress to be flopped into the bilge. To increase the comfort and security of on-water sleeping, a formfitted screened canopy is hung from the arch and stretched over the cockpit to the Lexan windshield, keeping out rain and critters—while a Nicro powered vent circulates fresh air through the coffin-like confines of the hull.

Another essential requirement is human-powered haulout—easy with a kayak but prohibitive for even small multihulls

MICROSHIP PHYSICAL SUBSTRATE

burdened by touring load. Since open-ended wandering precludes land-based trailers and tow vehicles, each boat carries four Seitech wheels... lovely 16″ diameter pneumatics with red plastic hubs and uncorrodable Delrin needle bearings rolling on a 1″ shaft. These deploy like aircraft landing gear from their nesting place under the solar panels, effecting the transition between retracted and extended states by pivoting around the planes of the crossbeam bulkheads via tensioned lines under control of levers at the gunwales (this complex but lightweight solution reluctantly adopted after much analysis of seductive but impossible hydraulic approaches). By dropping the wheels while approaching a ramp or beach, the boats can be dragged to a campsite or out of storm-tossed seas without logistical dependence on a chase vehicle.

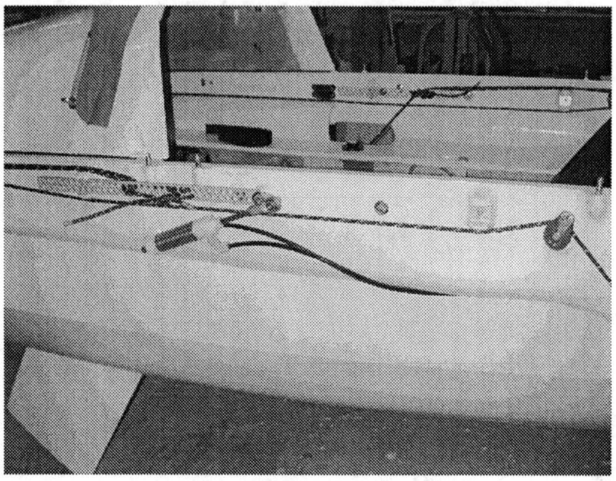

Landing-gear control levers.

Active steering of the forward landing gear turned out to be necessary, so there are hydraulic cylinders on each strut coupled through a stainless steel cable assembly to rotating collars, which in turn couple their rotation to the spindle bearings through scissor assemblies. To haul the boat, I simply strap on a harness and start pulling, with my hand on a winch handle protruding from a hydraulic system at the bow. This consists of a cam follower that implements the Ackerman steering function (to minimize wheel scrubbing, which is a non trivial problem), as well as an extra cylinder that biases the system in such a way that the front wheels become pigeon-toed... effectively holding the boat on a hill.

The landing gear suspension has 8″ of travel and can theoretically handle 4G shock loads, cushioning the impact of a hard surf landing with stacked bumpers of DuPont Hytrel from Miner Elastomer. (I sure hope we never get to test that feature!) The development of this system, with custom machined front-wheel spindle bearings, TIG welded tubing, deployment system, hydraulic steering, and countless unanticipated details, took over a year and many thousands of dollars... but it gives the amphibian Microships a capability unique in the boating world: unassisted launch and retrieval as long as there's a ramp or reasonably accessible beach available. A simple fixture built around the boathook plugs into the mast step and allows the 21-foot rig to be carried atop the folded boat, with the second support point being a foam block plugged onto the traveler on the arch.

Fitting crossbeams to the deck at structural bulkheads.

Much of the remaining "deckage" supports the fixtures necessary to meet the needs of our electronics systems—retractable solar panels and their waterproof interconnects, a thicket of antennas, video turret, data collection tools, and so on. Most notably, the control console of each boat is a sealed unit packed with computers and communication gear, bolted to a nest occupying the forward half of the eight-foot cockpit area and protected by a foam core cowling that carries the graceful lines of the boat much better than a funny box covered with hinged and gasketed panels. This was the most critical packaging problem, as anything else on board (other than the Macintosh laptop

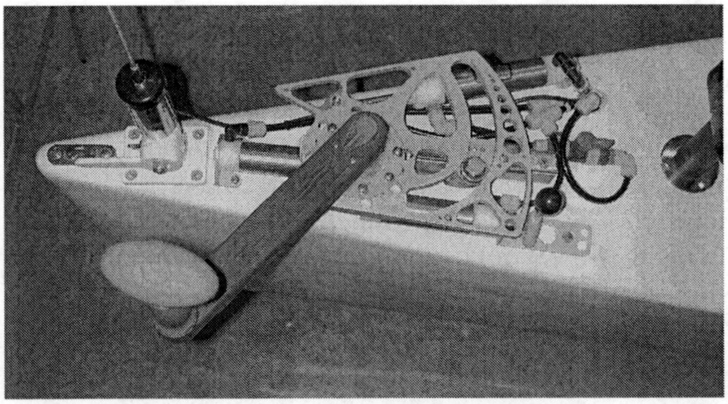

Bow hydraulic assembly for landing-gear control.

foam core panels support all the circuit boards, which are mounted on little standoffs screwed into flat pads that are simply glued in place.

Why foam core? The trick to this whole thing is not just careful sealing, but pressurization (at an estimated ½ PSI above ambient pressure), and the .062″ aluminum I used on *BEHEMOTH* would bulge like a balloon. It is impossible to achieve a perfect seal in something like this, and without running the whole box at elevated pressure, thermal cycling would result in constant breathing through the leaks… contaminating the interior with inhaled salt and moisture laden air. But in a pressurized box, the inevitable leaks just outgas, making them easier to find while allowing the control system to diagnose its own enclosure integrity by occasionally interrupting the pressure source and watching for a pressure drop uncorrelated with temperature change.

Canoeing just ain't what it used to be…

Forward retractable landing gear assembly showing hydraulic steering and suspension system.

nestled in a Pelican box in the aft hatch) can probably survive an encounter with salt water… but not the dense collection of surface mount printed circuit boards that satisfy our primal gizmological urges!

The basic skeleton of the console is a complex frame of molded fiberglass angles, filling the space defined by cowling, deck, and pedaling envelope as fully as possible. Every surface plane of the resulting micro-trideckahedron is surrounded by flanges on all sides, supporting hinged panels of thin foam core fiberglass. The exceptions are the baseplate, which is aluminum with stiffening ribs and welded mounting bosses for heavy hardware, and the "front panel," which is a single unbroken expanse of surface abrasion resistant polycarbonate behind which are mounted a dense array of LCD panels and other user displays. The

Completed cowling with hinges for access to internal electronics.

Propulsion and Power

The fundamental flavor of this project precludes nasty pollution-belching two-stroke outboard motors (which are fast becoming illegal anyway, and rightly so), but we obviously need auxiliary propulsion if we are not to be entirely at the mercy of fickle winds and currents. I mentioned the sail earlier—a 21-foot freestanding aluminum mast carrying a 93 square foot sock luff roller-furling vertically battened main, with an 8 foot boom on a gooseneck bearing to allow free mast rotation.

Bob Stuart and his SpinFin pedal drive unit.

A couple hundred miles on a Fulmar-19 with its deployable pedal drive unit convinced me: the way to human power a boat is with leg muscles, not arm powered paddles. A Microship can cruise about four knots with a small *SpinFin* drive unit made by Bob Stuart… faster than paddle powered kayak cruising speed despite considerably more weight and three times the number of hulls.

Each Microship carries one of these pedal thrusters, mounted outboard on the starboard side and retractable into a nacelle when not in use (or when it's necessary to access the prop through a hinged, gasketed access port for cleaning or replacement). Since the unit is side instead of center-mounted, the pedals are of the "double throw" variety; the SpinFin's input shaft is coupled to a removable machined assembly in the shape of a squarish 'S'. Removability is key—remember, we need to be able to sleep in the hulls

The thruster itself is an elegant piece of engineering. A ⅝″ stainless shaft on the left side of the device carries crank rotation and serves as a turning axis for deployment (via a lever inside the hull with adjustable latches and emergency kick up release). Molded to the end of the shaft is a stepped collar of filled epoxy and glass, carrying the roller bearings as well as a 90 tooth drive gear, coupled via ¼″ pitch chain to a 10 tooth cog on the propeller shaft. A tensioned idler reduces the cross section of the unit to the minimum necessary to pass the chain (which makes a 90 degree twist), and the prop is a slender 13″ two-blade unit. A molded skeg protects this from bottom contact, and the internals run in a light oil bath to keep it quiet and prevent corrosion.

The other auxiliary power source is solar – with each boat carrying a massive Solarex 480-watt folding photovoltaic array. Under optimum conditions, this can generate over 32 amps of 12V power, which is approximately full thrust for a Minn-Kota 42EX electric motor. Pivoting down when needed from the port afterdeck just behind the arch, this robust thruster is indexed into a steering motor that allows manual or computer control of propulsion angle, with hall-effect position feedback.

The solar panels themselves are a dominant feature of the finished boat—the size equivalent to nearly a ping pong table on each side, filling the spaces between hulls and overlapping the crossbeams. They are hinged to the center hull by locking pins emanating from the "gunwale bellies" next to the cowling, and are bungeed to inflatable cushions on the amas to allow flexion in response to cyclic heeling forces or the odd slamming wave. Nonskid tape covers all non active area to make moving around a

SpinFin in use.

bit safer, but the basic way to stroll between cockpit and dock is on the nonskid coated back of a folded panel—each side's array consists of four separate panels, hinged into two pairs. This not only allows retraction in rough conditions, but complete pack away removability for security when we're away from the boats.

One of the problems, of course, is that these things have to handle body weight—which, on a boat, can be amplified considerably by the dynamics of ungracefully falling down. To yield the best stiffness to weight ratio, we specified ultralight Solarex modules... fragile Tedlar sandwiches of photovoltaic cells without the usual aluminum or stainless backing. We then vacuum bagged fir framed substrates of one-inch Divinycell foam core, backed by ten-ounce fiberglass, and topped by a sandwich of 4mm okoume marine ply \and six-ounce satin glass. The solar panels themselves were then simply glued to these beefy structures which support hinges, latches, internal wiring channels, and all mounting hardware.

The electronic challenge here was to manage thruster power in such a way that we don't inadvertently kill the marine deep-cycle battery on each boat... cruising along at maximum power would, with no solar input at all, drain a fully charged battery in about three hours (defining a cruising range in swimming pool conditions of about 15 miles). Normally, of course, there will be some input from the solar array... as well as some battery charging current and additional load imposed by the systems in the console. Responsibility for intelligent power management is the domain of three embedded microprocessors developed by Tim Nolan on the Shipnet control network: every few seconds, they perform a complete power budget analysis, tweak battery charging levels, optimize the transfer of energy from the solar array using a peak-power tracking algorithm, calculate how much "free energy" is left over after all other needs are satisfied, then set that as maximum thrust at full throttle. It's thus easy to crank the throttle to full speed ahead and cruise on solar power alone, knowing that precious amps will be rationed in a way that won't deplete the batteries.

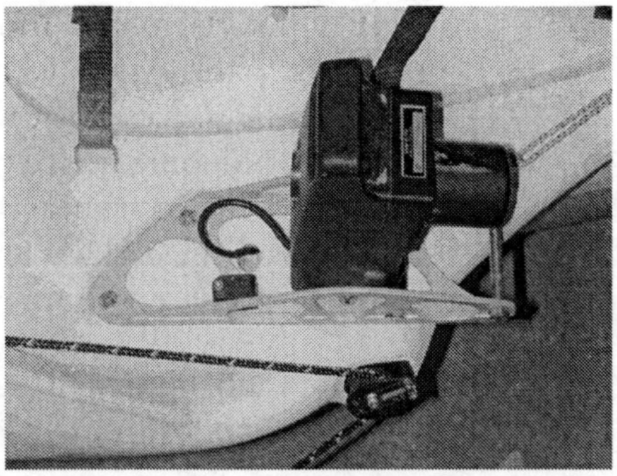

Electric thruster steering motor, retracted in its mounting frame.

(Naturally, there's an emergency override switch that applies full power to the thruster —if a freighter suddenly materializes on a collision course, the future status of the battery is no longer as interesting!)

And no, pedal, solar, and sail speeds do NOT add linearly... boats don't work that way. But it is nice to have options. Incidentally, we considered the seemingly obvious approach of integrating pedal and electric modes into a single system, but there is no good way to do it that does not involve jackshafts, clutches, speed reducers, and nonstandard hardware. Though it may seem redundant, having separate electric and pedal thrusters adds a lot of reliability and simplicity to the design. The combination of three propulsion modes gives the Microships maximum flexibility under a wide range of conditions without having to pollute the water, carry stinky fuels, and generate nasty noises.

Thruster control system development, with test tank.

The Microship User Interface

One of the primary design goals in this whole system is that ALL operations must be manageable easily from the cockpit. Not only are we constrained by basic single-handing requirements, but moving around on the deck can be fairly challenging under lively conditions—clambering over slippery solar panels to perch delicately on the bow while furling the sail constitutes flirtation with a potentially fatal man-overboard situation.

This has profoundly complicated the design, of course, for we're still dealing with a canoe hull here... there's not a lot of space to play with. The challenge (which is familiar to air- and spacecraft designers) is to integrate a huge array of controls and displays into the tiny space accessible to a human constrained to a seat.

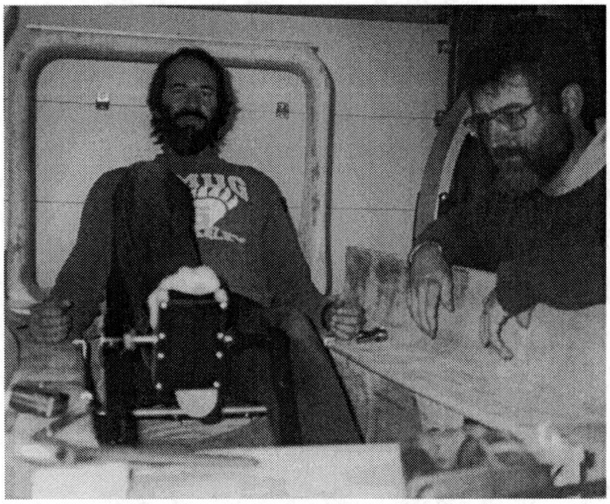

Seat, pedals and steering controls impose tight tolerances; here we're testing leg extension and pedaling envelope.

The Microship user interface tools fall into two broad categories: mechanical and electronic. The latter, with only minor exceptions, is contained within the console and a small cluster of controls; the former, a diverse collection of lines, levers, and hydraulics, introduces a complex set of physical design issues.

The cockpit was defined almost entirely by ergonomics—early on, we spent huge amounts of time in biometric analysis, measuring the operating envelopes of pedaling legs and reaching arms, while defining the balance between optimum eye point and overall height off the bilge. As it turned out, we had less than an inch of slack: the seating position is affected by boom clearance and visibility issues, which in turn constrained everything from the vertical dimension of the console screen window to the placement of rudder control handles. To allow fine tuning in response to seasonal clothing thickness variations and minor errors in our initial assumptions, the recumbent seat can be minutely adjusted using a vernier system to hundreds of positions and angles. As noted earlier, it is also retractable onto the afterdeck to allow sleeping within the cockpit.

In practice, everything ends up being defined by the pedaling axis; other than tweaking my slouch angle, there wasn't much we could do about my body dimensions. You can see how this became a rather circular problem—it took days of fiddling and measuring with kluged pedaling emulators to finalize the location of the pillow blocks and specify crank length, which, along with my foot dimensions, then defined the shape of the console's underside. Hot-glued cardboard models began to take shape, and once the seat was in place we could at last nail windshield dimensions, cowling shape, dodger placement, and the T-handle steering levers that I'll spend thousands of hours gripping at levels from

Bob Stuart slaving away on endless "deckage" details.

the delicate to the desperate. All other controls then had to fit within those dimensions, or at least not get in the way.

I should note here that the basic canoe cross section was modified considerably: we raised the gunwales a couple of inches, then added a deck that extends both inboard *and* outboard. On the inboard side, it provides a horizontal surface that is opened into a sort of coaming and supports the console; outboard, it curves upward into "gunwale bellies" that increase the perceived cockpit space, provide armrests, and support the edges of the solar panels. The resulting "decklets" also carry bearing cages for the pivoting T handles that are coupled to rudder hydraulics, while providing a more comfortable surface than the normal sharp gunwales of a canoe hull.

Now the challenge is to fit everything else in...

The vertical bellies turned out to be the perfect place for the four levers that effect landing gear deployment and retraction via low-stretched lines routed through a total of 18 turning blocks to the struts. The stresses here are occasionally huge, so we threw quite a bit of glass at it—with built-up latch assemblies on the inside and outside of both port and starboard sets to firmly lock the levers in either position. The inner downhaul handles are above my arms when in normal relaxed steering posture; the outer ones are accessed by unlocking and hinging up the solar panels. (Incidentally, those outer levers handle the uphaul of fore and aft landing gear, and not only carry the deployment lines but also hydraulic cylinders that rotate the forward wheels 75 degrees so they'll tuck nicely under the solar panels. This is handled on the aft wheels with bungees and offset line attachment, but the forward set had to be steerable, which complicated everything!)

Another lever in the cockpit takes care of pedal drive unit deployment; mounted on the starboard side under the deck, it locks in either position and handles the SpinFin via a thru-hull linkage. A rotating latch assembly with adjustable friction (based on a recycled drum sander mandrel) allows the drive unit to kick up instead of fracture in a grounding.

Across the cockpit from this assembly is the daggerboard trunk, and just above it under the port decklet is a linear hydraulic assembly with a rotating handle. This takes care of rudder deployment and retraction, lockable in either position—and again, the ability to kick up in a grounding thanks to a latch with a compressed rubber spring. All four appendages had to be designed with this problem in mind, adding complexity to the apparently simple uphaul/downhaul problem.

Bolting on the electric thruster mount during final assembly.

The bulkhead behind my seat is given over almost entirely to hydraulic system management. A manifold of ten valves allows any of the fluid circuits to be recalibrated on the fly, with a reservoir tee'd into the system to take up any slack. And speaking of stuff behind me, I do have to turn around to deploy and retract the electric thruster... there was just no good way to remote this. I also have to turn around to open the aft

THE MICROSHIP USER INTERFACE

gear hatch, but we must be willing to accept a certain level of inconvenience in a canoe!

So far we've covered the major mechanical devices in the cockpit, but we still haven't taken care of sail controls and other nautical essentials. This is actually a messy problem, because the tails of control lines can become extremely cluttered. The lines involved are:

- MAINSHEET, which enters the cowling on the port side via a turning block on the forward aka and terminates in a swivel cleat near the daggerboard. A little bungee fixture lets me hang the tail on the daggerboard trunk—this is the messiest one.
- FURLING LINES on the starboard side, terminating in jam cleats. This allows free rotation of the mast to set the sail area anywhere between zero and 100%. (At the lesser levels, sail shape is far from optimal; this is more a furler than a proper reefer.)
- TRAVELER CONTROLS on the arch overhead, a bit awkward to reach but at least they're out of the way.
- BOW PAINTER, which is stowed when not in use but must be readily accessible.
- DOCKLINES with eyes, likewise.
- ANCHOR LINE and corresponding pendant from the starboard ama to allow off-center anchoring off the stern via a bridle. The bulk of the rode is 185 feet of nylon webbing, spooled onto an "Ankarolina."
- SAIL OUTHAUL, which requires access to the boom (a pain when the dodger is in place).

The lines and their associated cleats and stowage tools add a layer of complexity to the cockpit, but there's still more... we also have to provide for a wide range of liveability components. There's a seven- liter water tank along with a small faucet and pump... a bilge pump against the bulkhead behind the seat... a magic hydraulic stopcock that locks the rudder... stowage for fabric dodger components not currently in use... and various nacelles for essential personal accessories, binoculars, hand bearing compass, old fashioned paper charts, sunglasses...

Finally, on top of all that, we have the electronics—waterproof speakers, microphone, wearable computer, sealed keyboard, pointing device... all those user interface goodies that connect me with the console system itself. We've been alluding to this box all along; I think it's time to take a look inside... but first, let me tell you about the day we first got 'er wet!

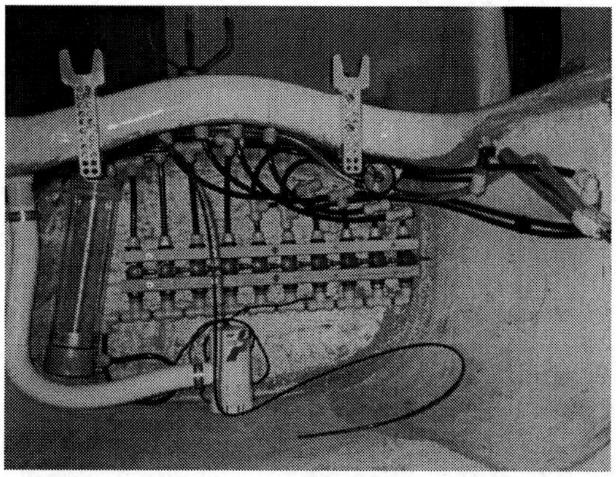

Hydraulic calibration manifold on the aft bulkhead allows easy tweaking of rudder and landing gear adjustments. Also shown is an automatic bilge pump.

First Microship Launch!

October 1998

Ah, now this is more like it. After an interminable epoch of being the Nomadic Research Labs facilities manager to bring the new building online, it is nothing short of exhilarating to return to the Microship project... especially with the recent thrill of the first ever test launch of the boat. We've been fantasizing about this for well over a year.

The mad push began about a month ago—you may recall my offhand mention in the last Nomadness posting that I'd be speaking at the West Coast Sea Kayak Symposium and planned to sail there. As in any industry where trade-show dates drive the development process, this became such a focused deadline for nautical functionality that an acronym was born to label all things that can wait: APT (After Port Townsend).

Parts were ordered; machining jobs specified; and lists written (then trimmed to reflect reality). What projects were absolutely essential to take the Microship on water for the first time, and of those, which would yield permanent fixtures and which would be test only? With the daggerboard trunk cast in glass, for example, the only knobs left to twiddle for helm balance were board length and rudder placement; we elected to kluge a nonretractable test rudder controlled by awkward bridle tugging rather than spend a month refining the hydraulics and kickup system for a real one that might end up in the wrong place. On the other hand, the adjustable seat, pedal drive system, rigging, furler, and console cowling would involve a huge amount of work whether we did them right or not... so we elected to go ahead and do them right. At this point, we kicked into high gear and maintained hacker's hours for the next three weeks (with a few all nighters thrown in for massive fiberglass layups and time-critical projects).

Tim Nolan, visiting on our Geek's Vacation program from Wisconsin, spent his first afternoon fashioning a temporary rudder fixture steered by a copper Tee plugged into the tiller hole, cheap cabinet hinges, and scrap plywood hot glued to rudder and hull. I spec'd a load of Ronstan rigging components—Series 38 ball bearing blocks, cleats, mainsheet traveler, and other nautical hardware—and got it all on order with only days to spare. Steve Goshorn, a machinist here on Camano Island, set to work fashioning a load of aluminum parts designed by Bob Stuart to allow a wide range of recumbent seat movement—with vernier adjustment of fore-aft, up-down, and angular positions. And it seemed that every other day we zoomed off to Everett or Anacortes to prowl marine supply and hardware stores for stainless bolts, pricey line, hatch straps and buckles, shackles, and essential bits forgotten on the previous day's trip.

Down in Silicon Valley, Andrew Letton started fabricating the pedals—a decidedly nontrivial project constrained by difficult stress problems, ability to withstand lateral grounding loads on the input shaft of Bob's SpinFin drive unit, removability for on water bivouac, and minimal operating envelope. After a few marathons and all nighters of his own, the package was shipped—joining the Ronstan goodies and other accumulated hardware on the bench with only a week until scheduled launch.

First Microship Launch!

Meanwhile, Bob was practicing his art—fiberglass and mechanical fabrication. He sculpted a beautiful cowling that defines the console shape, and drove us crazy with endless refinement of the foam core deck in preparation for painting (seemingly premature, but for the fact that ultraviolet rays damage unprotected epoxy resin). The results of all the filling and sanding marathons were worth it: once painted, the boat bespeaks smooth grace and balance with nary a glitch in her seductive curves. I fabricated a Divinycell and doorskin furling drum, then Natasha glassed it to the aluminum mast after first removing a ring of black anodizing and doing an acid etch to allow epoxy adhesion, whereupon Bob and I turned it down with the angle grinder to make it pretty. Little bosses and lands of filled epoxy were integrated into cowling, akas, and deck as needed to support various turning blocks, fairleads, and clamcleats—and the boat, for so many months a piebald patchwork of glass and goo, suddenly began sprouting shiny metal parts with a decidedly nautical flavor.

Final staging. A documentary film crew (Maureen, David, and Craig of *Picture This Productions*) arrived from Quebec, hoping to catch frenzied last-minute lab action and the uncertain excitement of first launch, and we decided to truck the boat to Port Townsend and do the test there—she just wasn't ready for a 50-mile sail. Bob assembled the drive unit and integrated it with the pedals using Andrew's machined Delrin pillow blocks and Trantorque couplers; I fiddled for hours with the critical drilling and mounting of precisely parallel seat supports on a deck with no flat surfaces, then moved on to rigging and the temporary SpinFin deployment lashup. It's all a sleepless blur in retrospect, but somehow we found ourselves slapping on a fast-drying Interlux primer/filler coat—even pressing the cameraman into service on the morning of departure to help finish the painting before first public rollout amidst a thousand kayakers.

In a mental fog, I piloted the mothership over 75 indirect miles to reach the Keystone ferry, where I paid $42 to take 'er aboard for the short hop over to Port Townsend.

Saturday dawned way too early, and before my normal latte time I was on stage chatting about electronics packaging methods on tiny boats. We extracted the Microship from the truck and rigged her on the workstand, perched incongruously atop tiny casters on the conference center's lawn. The furled 21-foot rig raked the sky; the prop spun smoothly with every touch of the pedals; the overall effect looked like the mutant offspring of a Klingon battle cruiser and a snowy egret. Crowds gathered and questions flew, the fascination with a pedal drive system evident in this paddling culture.

And then it was time.

Creaking and complaining, the workstand—never intended to be used outside the smooth concrete floor of the lab—began its journey. With four of us manhandling the akas and others forming a sort of protective phalanx, we rumbled through the soccer field and onto the roads of Fort Worden, creating a bit of a traffic jam and pausing at one point to heel the boat 30 degrees to sneak under power lines. At a busy intersection with a curious queue behind us, a cop strode over purposefully and opened his mouth to order me off the road, but in a flash David was there with the video camera in his face, asking what he thought of the Microship. He shrugged and waved us on.

After a long downhill reach to the beach, we dragged the tiny two inch wheels sideways to turn into the ramp area, paused a bit to strategize, then backed her straight into the water. The workstand by now was a bit wobbly and trapezoidal, with cracks radiating from drywall screws, but it held together until the boat floated lightly off and unseen hands wrestled it from the sea and stashed it on the beach.

And there she was... afloat for the first time, resting on the water as gracefully as a leaf even with an empty weight of 337 pounds! The design waterline painted on the canoe hull was parallel to the water; the amas barely kissed. So far, so good.

I clambered into the webbed recumbent seat (after ducking under an aka and soaking the crew's wireless microphone with salt water <cringe>), tightened the temporary steering bridle, deployed the thruster and locked it down with a ball detent pin into a temporary fixture on the hull, and slowly backpedaled... she effortlessly glided along the dock and turned to face open water. With a wave to my friends and the cameras, I kicked her into gear and she flew! Mike Cichanowski, president of Wenonah, paddled vigorously alongside in a kayak as I pedaled—giving me a useful data point on speed (I neglected to bring a GPS). Maintaining 3-4 knots was comfortable, amazing when you consider that the thrust is provided by a 13 inch, two-blade model airplane propeller... we'll gather hard performance data next time.

Once away from the docks and stray boats, I retracted the drive unit and drifted a moment until she was still. Time for the next test...

I uncleated the furling lines, which run 2.5 times around the drum and through a couple of exit blocks in the leading edge of the cowling, and wrestled with them a bit to find the sweet spot between slipping and binding. In moments, I had the sail unfurled, and rushed to tighten the outhaul running along the boom. All so unfamiliar, yet intimate: not only have I been specifying, building, buying, and staring at this stuff since the beginning, but there was also a tangible sense that it will all become very routine someday... even though the new inputs flooding my senses were so complex that I could barely keep up. The WindRider sail snapped into a lovely shape, instantly easing my fears about the vertical

battens, and I hurried to drop the daggerboard into the bizarre forward angled trunk and lock it down with <blush> a couple of tapered wooden shims left over from hanging my office door.

And in a rush we were off! Before I knew it, I was flying the windward hull, the canoe and leeward ama slicing through the water silently but with unmistakable speed. She tacked easily... and I was instantly relieved to see that there was just the slightest hint of weather helm: when I released the rudder, she would head up gently into the wind.

I was vaguely aware of Natasha chasing me around in a kayak with waterproof Pentax in hand, as well as David with his video camera perched in the front seat of a double paddled

FIRST MICROSHIP LAUNCH!

by Matthew Smith of Current Designs. Photos of the event are on our website... but at the time, I was lost in another reality entirely. The fantasies and design meetings, the PERT charts and ambitious specification documents, the late nights and countless dollars, the whole lab construction project... all snapped into focus and pointed as one to this moment. The boat actually works! Giddy with the sensations, I tacked back and forth in the waters off Point Wilson, realizing that the project had just made a quantum leap into its next phase.

Reluctantly, I refurled the sail and pedaled back to the dock—already growing comfortable with the turning radius and general feel. Bob took 'er for a spin under pedal drive only, pushing hard to test the system... then we adjusted the seat to give Natasha a go. It was all so surreal... I stood on the shore and watched this lovely little vessel scooting amidst the kayaks as sunset pinked the sky, while my mind, like Kai software, magically rendered the details of a matching boatlet, bristling antennas, sparkling blue solar arrays, white radomes, and LCDs alive with graphic reflections of my systems, my sensors, my life...

SYSTEM INFRASTRUCTURE

Despite severe space and weight constraints, the Microships carry an extensive network of electronic systems. As with *BEHEMOTH,* all of them can be loosely grouped into two broad overlapping categories: infrastructure and applications.

At the infrastructure level, the system has to deal with power management and distribution, thruster control, routing of audio/video/serial channels on demand, ship security, internal performance tracking and diagnostics, console pressure and temperature monitoring, dynamic network configuration including remote wireless control, and so on... the low-level tools that keep the boats working, provide access to resources, and are always on and thus must use power as efficiently as possible. In this category, we are jealous of every milliwatt, and this affects the design profoundly.

The applications layer includes obvious nautical systems such as navigation and communication, live satellite weather image display, scanning sonar imaging, digital video production and camera management, music composition and MIDI control, environmental data collection with hourly posting of telemetry blocks to the Net, general Internet access for email and web use, ham radio, manpack tracking, the whole suite of productivity tools accessible while sailing or off-boat, and ongoing software development.

It thus became obvious early in the development of the Microship system that some unique design challenges were present:

- Central server-based multitasking environment on each boat that can accept asynchronous data from a variety of sources and stream it to a variety of clients, including real time displays and a scriptable database...
- Random software controlled interconnection of analog and serial channels...
- Distribution of independent control resources across a low-level network to simplify development and minimize power drain...
- Graphic front-end access via web browsers, whether local to a boat or via wireless remote connections...

Let's take these key features one at a time...

ON-BOARD SERVER

I should have expected this. When I started working on the Microship project back in 1993, uncertainty about how to proceed with the nautical substrate induced me to focus instead on the control network. With the help of UCSD students and industry volunteers, a system took shape: a FORTH 68HC11-based multidrop network. It actually worked very well, and we started writing front end tools—briefly in C on a PC, then in HyperCard on a Mac, then in NewtonScript to support wireless handheld computers. I vaguely recall joking about how it would all be obsolete by the time I finally got the boat done, and, well, that turned out to be the case.

Obsolescence is a touchy subject... some people abandon old systems just because marginally faster new ones are available; others hold on to once whizzy clunkers until antique value causes the price to start back up. In our case, the deciding factor was a sea change in networking technology and the availability of single board computers so robust that we could run a full scale linux server with modern tools... for about the power cost of a typical nightlight. What does that buy us? Nothing less than the ability to see the Microship as a web server, with all control and monitoring widgets implemented as a website, available both locally on the boat's console screen or from any computer within a few miles... or anywhere in the world, for that matter, given our satellite Internet link.

Phew.

Using a single board Pentium-class computer from Octagon and a heavy infusion of software wizardry from Ned Konz,

SYSTEM INFRASTRUCTURE

we brought up an implementation of Debian GNU/linux and went to work. Throughout this process, we redefined the intent of this whole system, eventually viewing it as a set of processes clustered around a central database and communicating through sockets. Each physical resource is "owned" by a single process, such as the Hub server that controls the connection with the long established New Micros FORTH board that (together with a boatload of interface hardware) manages the crossbar networks and handles all the low level I/O on the ship. Any process that wants to talk to hardware, whether to schedule an audio crossbar connection or snag a snippet of environmental data from a sensor, thus becomes a client of the Hub server, which in turn handles the gritty details of interacting with the FORTH interpreter and other low-level tools.

One of the ongoing objectives here is the collection of many flavors of data pouring in through serial, analog, and digital channels. Readings are time stamped and dropped into a database, multichannel snapshots are transmitted to the public web server on an hourly basis, equipment performance datasets are archived for sponsors, and—depending on the current situation—any arbitrary set of inputs can be displayed live on the console browser as stripcharts or simulated instruments. A key part of all this is the Realtime Data Broadcaster, which provides a publish/subscribe service for any client interested in receiving realtime data.

The fun part about all this, of course, is the variety of applications that now become possible. An example is the virtual console that will run on the canoe's browser: it's a configurable screenload of applets including a scrolling chart segment with a current location icon, basic nautical navigation tools, relevant live sensor data, any subset of eight moving strip charts vertically arranged on the same timebase, power system performance instrumentation, and links to everything from extensive online documentation to close-up windows focused on individual subsystems. Bear in mind that this architecture will be replicated on Natasha's boat, and possibly even in the manpack computers… allowing any browser in our wireless network to connect to any of the local URLs just like surfing to a website on the Internet, yielding a fully capable front-end control panel complete with online documentation, deep archives, image feeds, live audio, real time displays, messaging, and more.

Most of this high-level stuff is written in Object Perl[1], and takes advantage of the whole industry that has grown up around the Web as well as the open source linux community. But the real beauty of this application depends on interactions among distributed resources, requiring quite a bit of low level code and hardware hackage…

GRAND CENTRAL STATION

One of the most troublesome problems facing this design was the sheer number of hardware devices that have to interact. We're not talking bus-compliant systems speaking a standard interoperable data format here; we're talking about celphones, marine radios, video cameras, sensors spewing NMEA or RS-232 datastreams, frame grabbers, tiny dedicated control interfaces and translators, and so on. From bike experience, I know that any "complete set" of anticipated interconnects will become obsolete the first time a seductive new gadget arrives—the *Winnebiko II,* by the end of its era in 1988, was a study in reassigned console switches and multiple layers of annotation scrawled upon already lousy documentation.

For *BEHEMOTH,* we developed an audio crossbar around the Mitel 8816 chip, vastly simplifying those spur-of-the-moment connections that are irresistibly associated with having lots of devices. It took only a few lines of FORTH code, for example, to respond to a lat-long change without the right password by dialing 911 on the cellular phone and having the speech synthesizer intone, "I am a bicycle and I am being stolen. My present coordinates are…"

This sold me on the crossbar architecture for like-flavored signal paths, and the "Grand Central Station" region of the Microship system includes three of them… a cabling-intensive zone in the console with 244 connectors.

The audio crossbar (Auxbar) allows up to eight simultaneous line-level connections among any of 32 sources and 32 destinations. This consists of a stack of two boards bristling with gold plated RCA jacks, so all we have to do to add an audio device is run a cable and add a line to a config file to allow reference by name. Natasha's boat joins the network via a short range stereo full-duplex transceiver for quality audio and a 70 cm radio link for voice grade exchanges… devices which are, of course, merely addresses on the Auxbar. At her end, a scaled down version of the same design allows the addition of multiple devices as well.

The video crossbar (Vixbar) is similar, but optimized for higher bandwidth signals and the need to channel switch during retrace intervals to prevent glitches. This system can handle eight simultaneous links among any of 16 sources and eight sinks, and, since it is less likely to be on call 24 hours a day, is supported by a dedicated FORTH node and software controlled power switch. This network is also expanded to the other boat via a video transceiver, since Natasha runs the Draco Casablanca video editing system and needs to select camera channels on the fly; like audio, this system appears as an array of RCA jacks and is trivial to expand or edit. One-line commands make or break any link, add a receiver to an existing connection, or clear all of them.

The serial crossbar (Sexbar, of course), provides up to four simultaneous bidirectional connections among any of 32 serial channels—28 of which appear as a dense 4x7 array of DB-9 connectors pop-riveted together. This one is particularly amusing, since it also has to solve the annoying problem of RS-232 polarity, those eternally reversed pins 2 and 3. The code in the Sexbar controller, whenever asked to establish a connection, first zips through the involved pins of both channels, connects them one at a time to a window comparator, sees who's receiving and who's

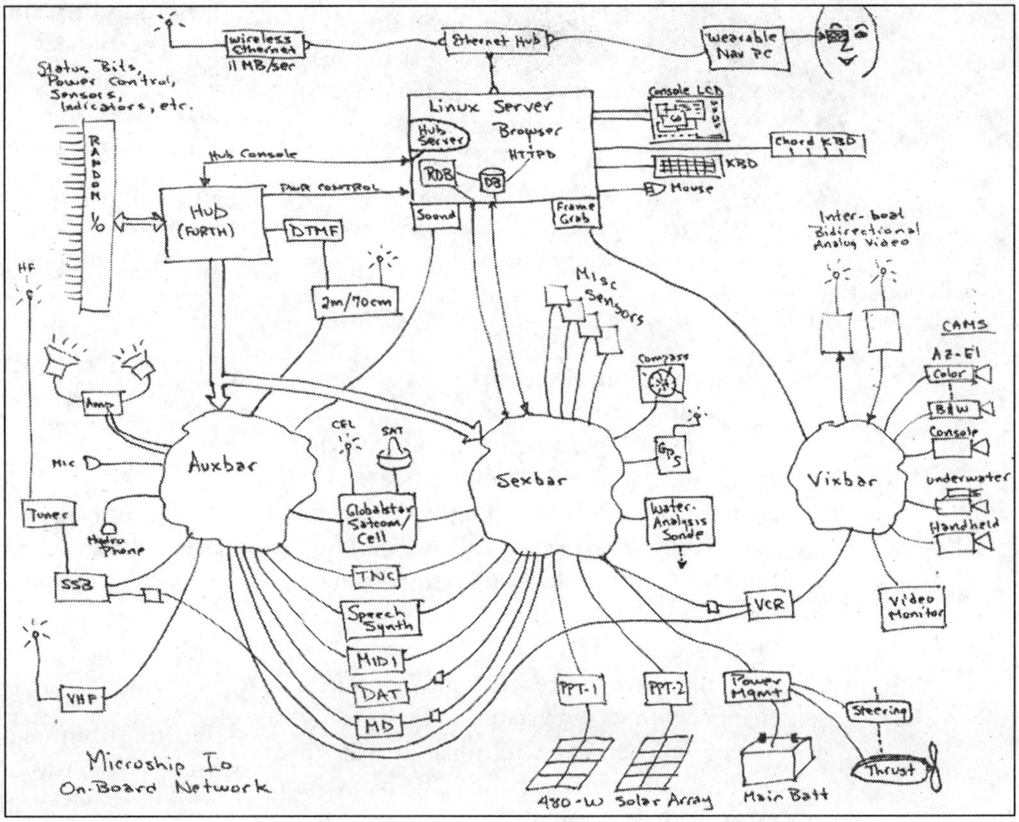

Conceptual sketch of Microship resource network, depending heavily on crossbars and a graphic front-end server.

System Infrastructure

transmitting, then assembles a virtual straight cable or a virtual null modem cable as needed. The process is so quick and painless that we often link serial gadgets together just for the fun of it, a statement I haven't been able to make since, oh, about 1973 when the process of making machines communicate was still intrinsically astonishing and UAR/Ts were radical new alternatives do doing it with TTL shift registers and control logic. <creak>

I should address one issue before continuing: I occasionally take flak in these Internet-savvy days for using such primitive networking tools as vanilla serial channels loping along at a poky 9600 baud via hardware-intensive crossbar switches. Why not just give everything a TCP/IP stack and do it right? Well, if the involved devices were all "real computers" that would be sensible, but most of the serial widgets we're using are small, low power, dedicated controllers... even PIC processors acting as translators between RS-232 and proprietary protocols like Sony LANC and Dallas MicroLAN. Sticking them all on Ethernet would be power-hungry overkill, adding far more hardware overhead than the devices themselves!

Grand Central Station: Auxbar (top), Sexbar (lower left) and Vixbar.

The power control system deserves a quick comment in this section even though it's not a network, per se. In keeping with the spirit of maximum flexibility, all devices that are even remotely power hungry are hanging on solid state relays, in turn wired to Phoenix Contact blocks that carry a few dozen bits of parallel I/O attached to the Hub processor. It is a trivial process (running one wire) to add a power control channel with its own status LED, fuse, and pair of switched screw terminals; it takes just another one-line command to turn anything on either boat on or off.

That takes care of the interconnect layer of our Microship network: with very few exceptions, all random pieces of communication, entertainment, data collection, security, and control hardware appear to the server as nothing more than some set of audio, video, serial, and power channel names.

The Hub and Other Stray Processors

An interesting problem confronts the designer of a power-limited portable system —even with 480 watts of solar panels and a big honkin' marine deep cycle battery, we still care deeply about how many milliwatts are being continuously converted into heat. The linux server is designed to go into a succession of power saving modes (slow clock, disk off, suspend, shutdown, etc.) depending on activity, but even when we're away from the boats and in no need of graphic front end tools we still want a faithful electronic watchdog keeping watch over power, security, bilge pump activity, and the like. Besides, some system has to be alive enough to respond to a call, transmit location beacons... and boot up the big iron if necessary.

In addition, there are so many little distributed tasks around the ship that trying to do it all with one central computer would create an I/O nightmare, a software quagmire, and catastrophic single point failure potential. Even with Mimsy (the web server) shut down, the Microship needs to be capable of taking care of herself in a surprising variety of ways.

The solution to all this is a distributed network of processors engaged in a variety of tasks—autonomous control systems that are happy on their own but can be queried or controlled by server-level code when the big system is alive.

The most central of these "little guys" is the Hub, which was once in the exalted role of running a whole multidrop network of nodes in the early years of Microship design.

This folding board stack is based on a New Micros 68HC11 with FORTH in ROM, with loads of added I/O including 64 bits each of individually addressable input and output bits along with analog and serial ports. The packaging includes a hinged "nexus board" that simplifies random little interface hacks, and the system also controls Auxbar and Sexbar. A custom multitasker by Bill Muench adds considerable utility, as monitoring or control tasks can be handed off to the board and then forgotten, with the results accessible through an easy textual exchange of variables via the Hub's console port.

(As mentioned earlier, there is a Hub server process that runs under linux, converting FORTH's vanilla serial console port into an internet-domain socket with full resource locking. We can open multiple telnet sessions to the little eight bitter, for example, and bang away asynchronously without difficulty. It's very strange, telnetting to a chip via Ethernet...)

The Hub stack, showing nexus board and I/O headers.

Anyway, the Hub's role in the Microship is both central and subservient: it handles all the signal routing on demand while keeping an eye on all the security and low-level monitoring jobs, reporting to Mimsy as needed (even yanking hardware lines to boot it up if something needs high level attention, like sending an email message or going into full-scale alert mode). This turns out to be quite convenient, for one of the Hub's many local I/O devices is a DTMF (Touch-Tone) decoder that sits on the audio crossbar. With the boat completely shut down and parked miles away, I can still send a sequence of tone commands from a ham radio transceiver or cellular phone, causing the Hub to wake up the server so I can surf my way over to the full scale user interface. And the Hub is empowered to beep me and sound basic security alerts even if it's all alone on watch.

There are other baby micros scattered around the boats as well, of varying robustness. All of them communicate via vanilla serial ports through the Sexbar, not only with the server but with each other or even remote users... useful, for example, when one of our software developers wants to log in and hack <shudder>. Following is a sampling of embedded processors aboard the Microships, some within commercial products, others developed to solve specific problems. From the system perspective, these are all simply serial devices that either spew data continuously or respond to queries with requested variables or some kind of action...

- Peak Power Trackers, continuously optimizing eight 60-watt "channels" of incoming solar power and delivering detailed performance data on demand.
- Thruster and Power Manager, which keeps an eye on the battery and all system loads, tells the motor's PWM controller how much free power it can have, and takes care of software-controlled thruster steering with Hall-effect position feedback.
- Video Turret Controller, which aims either the zoomable color or low-light monochrome camera at any specified angle, or scans at any rate between any pair of angles. This takes care of camera power control, lens shield servo retraction, and turret internal environmental monitoring.
- Video Crossbar Controller, which for historical reasons happened to end up in a dedicated processor.

SYSTEM INFRASTRUCTURE

- Wind Sensor, an ultrasonic marvel from Handar on the stern that deuces wind speed and direction by resolving differences in the speed of reciprocal sound propagation among three transducers.
- Electronic Compass, continuously transmitting heading information at the rate of one sample per second.
- GPS Receiver, which provides a continuous feed of latitude, longitude, and course data.
- APRS and Packet Nodes, which take care of location telemetry among our wandering group as well as email with the ham radio community and the transmission of periodic tracking updates to public map servers.
- Virtual Front Panel, a little PIC-based circuit that maps 48 arbitrary bit conditions onto a front panel LED matrix.
- HF Radio Controller, responding to serial commands by tuning and otherwise tweaking the all-band amateur radio transceiver.
- Water Quality Analyzer, a remarkable multichannel probe that returns massively detailed reports about the stuff we're floating in.
- Random Controllers, which translate serial commands into the proprietary interface protocols embedded in the DAT, VCR, and MD devices.
- Speech Synthesizer, a standalone hardware text-to-speech board that lets any device in the system have its say (the linux box has its own software synthesis using ViaVoice).
- Wee Embedded Sensors scattered all over the boat, single-chip parts that input analog values representing temperatures, voltages, stresses, or other parameters and translate them into continuous serial data streams. Others of the same class convert serial commands into local control bits.

And so on. As you can imagine, these are all so thoroughly different (and from a networking perspective, so wimpy) that we'd have a real mess on our hands without the Sexbar. With it, putting this wide range of resources to work is a relatively simple matter of writing drivers and filtering incoming data streams (with parsing hacks and regular expressions) into values that can be retransmitted by the realtime data broadcaster to webpages, databases, navigation software, control loops, monitoring processes, or even little daemons that periodically pack up a telemetry block and email it via satellite to somewhere far, far away...

FRONT END AND MIMSY

So how does this all translate into the day-to-day life of traveling on tiny boatlets? If I'm strolling around town with Microship *Io* hauled up on a beach, it's nice to be notified of security alerts, check power system status, confirm my location relative to the boat, and so on. Suppose I want an offshore weather update, delivered to my pocket walkie talkie in a synthesized voice? The ease of interconnecting resources turns the following into a one-click operation in a web browser or even a spoken command: power up the Icom HF single sideband radio, tune it to the Coast Guard NAVTEX frequency for digital weather report transmission, pipe received audio to a packet radio TNC, send the resulting data stream to a speech synthesizer, then direct its voice output to a radio transmitter. This was one of the first tests of the completed crosspoint network, in fact, and writing the software to do it took about five minutes (most of which was looking up or assigning crossbar channels!). After that, it was a single click on a button, displayed on a wireless handheld computer.

This top-level server running the network is Mimsy—the Microship Information Management System. Seen by the user as an on-board web site, Mimsy pulls together everything into a single browsable environment: our huge library of system hardware and software documentation, all databases, navigation and charting tools, calculation and engineering widgets, reference material, a copy of our publicly accessible www.microship.com web server, text archives, photo libraries, video channels, and the interactive control network itself.

The control system is certainly the most amusing part of all this, as it presents live data from the low-level network while allowing direct intervention in system activity. Surfed locally through Netscape, there are tools for everything: setting up communication events, collecting and graphically presenting environmental data, monitoring or overriding the thruster management software, controlling the video turret, setting up custom security or watch functions, programming the autopilot, or whatever. All this takes place via any web browser on the Ethernet… and since the two boats and two backpacks are all linked via high-speed wireless networks, this means that all resources are available all the time (even from thousands of miles away via password-protected cellular or satellite link).

In addition to presenting all this in a single smooth environment, Mimsy busies itself with collecting both internal and environmental data and sending telemetry blocks via email to our land-based server on a scheduled basis. It continuously monitors the health of all low-level systems and restarts crashed nodes, reporting such events to me via the console display but otherwise relieving me of the need to keep track of gritty internal details. It archives everything in a sprawling database, yielding an annotated breadcrumb trail of the entire adventure with attached photos, time and location stamps, and full system telemetry. Mimsy is the central tool that is used to manage the whole project (and, as such, our lives).

Of course, this system can't quite handle everything—certain functions must exist in parallel with Mimsy and its subordinates. Navigation and charting tools, in particular, require a certain degree of autonomy, and after much agonizing we decided to install a separate nav PC in the form of a wearable computer (the heads-up display is particularly well suited to navigating in the fog, especially if a head mounted compass is used to re-orient the display to overlay a computer generated wireframe model of reality onto the soup). Since so many excellent off-the-shelf nautical applications are PC-based, this is also the machine that handles live weather satellite images, phased array sonar displays, tide and current info, and so on.

In addition to those system displays, the console hosts a suite of dedicated instruments: marine VHF, electronic compass, standalone GPS, battery monitor, and so on. As much as possible, other display-intensive subsystems are integrated completely into Mimsy and the Nav PC. After all, this is still a canoe… we're hurting for space!

Finally, the user interface to all this is of critical interest, since we wish to avoid the clutter of hard-to-seal and uneditable dedicated switches and controls. The VHF, compass, navlights, and GPS are standalone for good reason, since if all the fancy stuff crashes we still need those basics to make landfall safely, but for the most part everything else is integrated into the high-level system—controlled by a chord keyboard on the right-hand hydraulic rudder control, a floating sealed QWERTY unit that tucks behind the daggerboard trunk, and a sealed torque-sensing pointing device for mouse functions. The left rudder control is integrated with a throttle for the electric thruster as well as push to talk for the currently selected radio, and a small switch panel on the left "wing" of the console controls navlights and other hard-wired basics.

[1]As this publication went to press, we were starting to consder doing the whole package in Squeak, a small talk implementation.

A Sampler of On-Board Applications

It occurs to me as I near the end of this lengthy tale about the Microship project that the massive amount of on-board technology may, at first glance, seem a bit excessive. I mean, even with the luxury of a graphic front end, there's only so much low-level control that you can do on what is basically a canoe with training wheels. Beyond the obvious marine navigation and fundamental technomadic needs, what might we want to actually DO with all this?

So before we leave our little peek under the hood, I'd like to give you a quick look at some of our other applications for this boatload o' goodies...

System Diagnostics

One of the original motives for the whole data display system is simply being able to keep track of what's going on in the system itself. One could argue that by eliminating the system, I would also eliminate the need to keep an eye on the system... thus knocking a year or two off the project and cutting weight by a hundred pounds or so. But a big part of this is the underlying geek expressionism that requires a dose of complexity in the midst of languid escape from the madness of civilization.

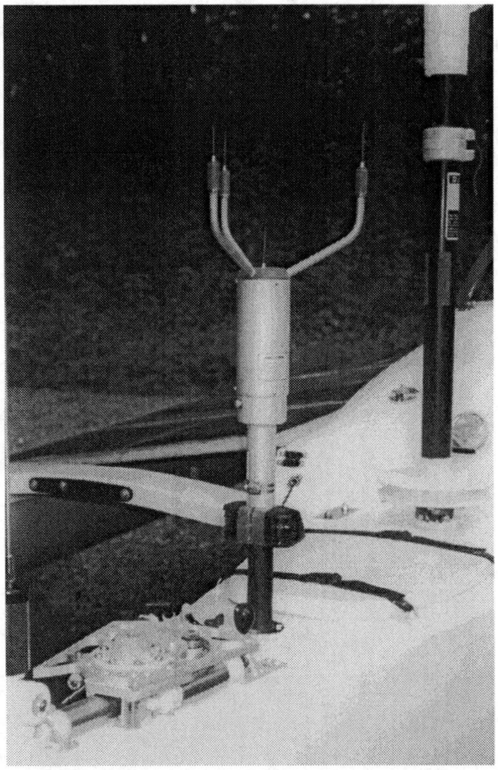

Handar ultrasonic wind sensor.

The physical Microship components generally don't require much in the way of instrumentation—I can clearly see and feel them, and on a boat this small, visceral seat-of-the-pants feedback is probably a better indicator of sail trim than, say, microbending lossy optical fibers embedded in the mast and strain gages on the blocks to report stress levels. But there are lots of things I can't see that are just as critical.

Take the solar panels, for example—a vast array for such a small craft, totaling sixteen 30-watt modules bonded to eight vacuum-bagged foam core panels, hinged into four assemblies that fill the spaces between the hulls. With the peak power trackers extracting some 15% performance improvement, we're expecting as much as 36 amps (at 12V) under ideal conditions. But all sorts of things will reduce this ideal number: dirt, insolation angle, elevated temperature, shading, saltwater incursion, wiring failures, and tracker demise. As such, the eight channels of 60-watts each are continuously reporting their performance to both a standalone diagnostic LCD and the server, viewable as real time stripcharts on the Microship solar array web page... little Java sandboxes that fill with data from the database (depending on selected range), then scroll new data into the graphs from right to left. This not only lets me view them all in a stack to observe a possible problem, but also explore historical performance and extract statistical information... as can Tim Nolan, the power management guru who designed the peak power trackers, logging in from his home in Wisconsin.

This same idea extends to other subsystems: console pressurization and sealing, thruster control, humidity inside hatches, bilge pump cycling, internal power distribution, system reboots, and so on. Being able to collect and manipulate data makes a wide variety of diagnostic tools fairly trivial to implement.

Environmental Data Collection

In a rather similar vein, from the perspective of the systems involved, we are taking advantage of on-board technology to do our small part to help clean up polluted waters. With years of anticipated coastal/inland on-water time and the technological tools on hand to collect and transmit environmental data, it would seem almost negligent not to do so!

We have no particular political or scientific agenda, and are thus gathering a wide range of easy-to-capture telemetry data. Some of it is purely internal as noted above, and some is navigational (speed, GPS coordinates, heading, and the like). But quite a bit falls into the broad category of environmental measurement:

- Air temperature
- Ambient light
- Barometric Pressure
- Dissolved oxygen
- Humidity
- Ozone
- Radiation (both air and water)
- Ultraviolet intensity
- Water pH
- Water salinity
- Water temperature
- Water turbidity
- Wind direction
- Wind speed

We are seeking other sensors that easily yield relevant and interesting data, with particular interest in hydrocarbons and ion-specific indicators of common industrial and agricultural pollutants (we're seriously lusting after a YSI Sonde and a Voyager portable Gas Chromatograph from Perkin-Elmer). Although we're not doing "real science" in the formal transect 'n survey sense, we do have the opportunity to collect location-stamped telemetry blocks at any interval and transmit them in real time to a publicly accessible web browser. The resulting river profiles should yield interesting data as we pass industrial areas, population centers, and regions of heavy agricultural runoff... and to make it easy to use, we are building a server-resident database to host the raw data along with a set of front-end tools that allow graphic observation of any group of channels over time—including the ability to click anywhere on our "breadcrumb trail" for a snapshot of the values at that point accompanied by corresponding observational text and images.

Graphic presentation is the first step in making raw data interesting, and—with the added rich context of the Microship adventure—we are targeting this system toward schools. To really deliver something of value (one-pass data is too much affected by rains and seasonal variations to allow more than casual conclusions except in obvious point-source pollution cases), we must go beyond the data itself and port the entire set of hardware and software tools to our educational clients. Students who grow enamored of nomadic data collection by watching graphic sensor logs of our cruise downriver can copy the designs, download the software, and apply the same techniques to local environmental problems. Such continuous monitoring over an extended period can indeed yield real science, not to mention a firm grasp of principles ranging from biology to wireless networking. These designs will allow relatively non-technical folks to build unattended monitoring systems that

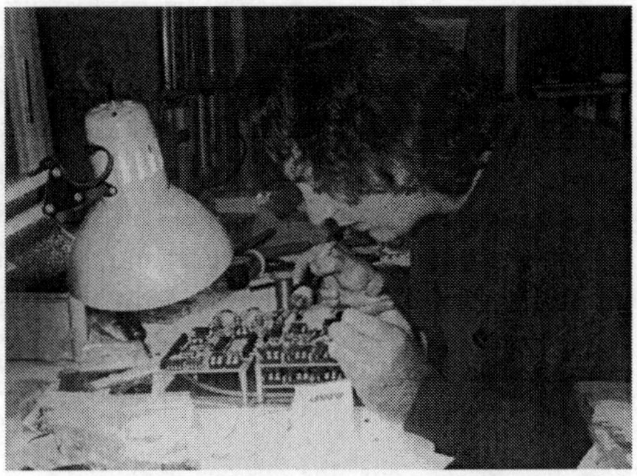

Tim Nolan at work on the eight-channel peak-power tracking system.

A Sampler of On-Board Applications

communicate with classroom computers via a variety of data links, inviting a whole set of potentially interesting student projects and collaboration with the scientific community.

Along these lines, we might present the Microships to schools as a sort of "experiment bus" along the lines of the Space Shuttle (but much more likely to drop in for a classroom visit!), perhaps with a competition that would encourage students to design a new experiment or sensor channel that we could add to the system. Coupled with a series of on-site school appearances while underway, this should make the whole affair quite engaging as an educational project; during the BEHEMOTH era, I delighted in getting kids turned on to technology through something "cool"… a much-needed alternative to dry textbooks.

Communications

This is a pretty broad topic, but justifies a bit of commentary since it affects just about everything else on board. Much of what we are talking about depends on a variety of local and global voice and data links, and on a vessel this size it becomes a real problem to package all the radios and antennas without creating a quagmire of interference problems. Processor boards spewing broadband RFI are enough of a problem…

Video turret, shown without protective acrylic cylinder.

Let's take a tally of the radio links we need. On the local level, there are a number of communication paths connecting *Io* and *Europa* with each other and the backpacks: wireless Ethernet for high-speed data and a suite of IP services, the "Packnet" that lets low-level guys chat without involving the big iron, amateur dual band HT's and license-free FRS walkie talkies for casual person-to-person chitchat, and high-quality bidirectional video and audio links to extend the crossbar networks across both boatlets with minimal insertion loss. All these are relatively short range links, and at the rate things are going, we might find that they all become virtual circuits through a single high speed spread spectrum data link. (The pace of technological change is why we wait until the last minute to acquire computers, wireless data hardware, and nav tools.)

When we start to look outside our little flotilla, a new set of tools is necessary, although there is some overlap. A key voice accessory is a handheld dual band ham radio transceiver for line-of-sight communication with locals and access to repeaters, and the data equivalent is a minor reconfiguration of the existing packet radio link to connect with local gateways. This is the simplest tool for our telemetry transmission—just drop email into the amateur radio store-and-forward packet network. In the boonies, with no local repeaters or digipeaters, we'll switch to HF PACTOR… as well as APRS, the position-reporting protocol layer atop amateur packet radio.

This is neat stuff. APRS appears in the system as a small box that automatically transmits position reports between our boats and packs at any preset interval. But it's not just local—a worldwide network on 144.39 and 10.151 MHz collects and forwards all such transmissions to a central Internet-delivered data stream, allowing our location at any moment to be viewed on a map through any web browser. It's all quite magical—have a look at http://map.aprs.net/microship, and if we have transmitted a position report in the past eight hours you'll see us with 500′ precision in three successively zoomed MapBlast windows. Ahhh, technology…

APRS offers some capability for additional telemetry channels, and we'll transmit a subset of our data stream through this path. But the full blocks will consist of 50 or so channels, and will be accompanied by current text reports, embedded notes for our home base, and still-frame image data. One of the linux server's jobs is to assemble these automatically generated bits of email, then choose the most appropriate path based on cost, location, and content. The choices at the moment are amateur packet via VHF or HF, cellular modem, or Globalstar low earth orbit satellites. The latter is a fully operational worldwide commercial network, and is expected to become the backbone of our wide area network connection with the rest of the world.

As wireless network infrastructure grows, we'll be experimenting with spread spectrum radios operating under both FCC part 15 (license free) and part 95 (ham radio, with business traffic prohibited). This technology is in a state of flux, but our goal is clear: maximum bandwidth at the lowest financial and power cost, regardless of location. We'll be depending on this not only for telemetry and email, but also to support remote software development and system administration, security when we're away from the Microships, server backups for essential files, website maintenance, and everything else that has moved Internet access squarely into the middle of Maslow's hierarchy of needs.

Video Production

One of the more whizzy and exciting components of the Microship project is something that I always wanted with *BEHEMOTH:* video. Lots of video. Cameras and hydrophones underwater, cameras on deck. Cameras in the consoles, staring at our faces. Even cameras that can be remotely steered from afar…

The most amusing video source, in fact, is a video turret on *Europa's* arch, operating under software control as a node on Natasha's Shipnet. Two cameras—a high-quality zoomable color unit and an infrared/BW low-light microvideo unit—are mounted on a rotating platform that can be quickly directed at any angle, or set to scan at any speed between any two angles. This has obvious value as part of the security system, and can unobtrusively gaze at any scene of interest, even under the control of a web browser thousands of miles away. On *Io,* similar functionality with somewhat lower video quality is accomplished by adding a small color/BW/IR/laser pod atop an az-el remote controlled searchlight with added position feedback.

The Vixbar (video crossbar) integrates all these sources across both boats, both consoles carry monitors, and a software-controlled 8mm video cassette recorder can record any channel. There's even a video transmitter that can send any requested channel to a pocket TV, though we may toss that now that wireless bandwidth is getting to the point that we can just pipe compressed video streams to PCMCIA frame grabbers in our backpack laptops.

But the sexiest part of all this is the production system—a nonlinear digital video editor known as the Draco Casablanca. This is integrated into Natasha's console, allowing her to work from any source material (including digital video cameras and computer generated images), manipulate clips with frame-precise cuts, assemble a production with a storyboard, generate hundreds of different transitions and titling effects, mix voice-over and music tracks, and output the whole finished work with broadcast quality… all from the confines of a canoe hull. An internal Quantum 18 Gigabyte hard drive can store nearly an hour and a half of professional quality video, and these are easily swappable to allow her to work on multiple projects.

While we travel, we'll produce video tapes of our adventures, the technology that makes it all possible, and amusing encounters along the way—a major upgrade to the text-only storytelling that has always supported my technomadic adventures.

Stay tuned!

A Sampler of On-Board Applications

Music

While we're on the audiovisual subject, I should mention another key application that is being designed into the Microship system. Actually, it's two applications...

First, it goes without saying that a decent stereo and music library is essential, and it's amusing to recall just how close I got to buying a Sony 150-CD changer and interfacing it to provide a mobile music library. I'm sure glad this whole gig is coming together at the turn of the century, for now we have MP3 instead—the music compression algorithm that has swept the Internet and caused the recording industry more than a little panic. With roughly 10:1 compression ratios, CD-quality audio can occupy about one megabyte of disk space per minute of playing time... or 16 hours per gigabyte. At the vanishing cost of hard drives these days, there's no longer any reason to carry those bulky CD's that were such a key resource on *BEHEMOTH.*

In fact, experiments here have demonstrated that only one of us needs to tie up the disk space, since streaming a megabyte/minute via wireless Ethernet is trivial. The central linux jukebox can easily support two users listening to different material simultaneously while going about its other jobs unperturbed.

And speaking of high-efficiency music storage, there's another system on board that should offer endless amusement: a full-scale music studio consisting of a KORG X5DR MIDI synthesizer module, sequencer software, multi effects processor, mixer, and recording tools. This little ensemble will not only play the virtually infinite variety of .mid files on the Net, but also provides a computer-generated backup band to make my improvisational jazz and blues flute playing more interesting (I get in a rut when playing solo). Naturally, all these components are interconnected via the crossbar networks, so any resource combination is only a quick script command away... like generating music tracks for video production, adding amusing effects to the local response to security violations, and having endlessly interconnectable toys to while away the hours of a long crossing.

Security

Finally, I should say a few words about one of the more compelling "high level" applications—keeping the boatlets safe and, equally important, reducing our paranoia index in a wide variety of situations. The best lock is the human eye, but I know from long bike experience that being tied to the machines when there other interesting things to do gets boring.

In my travels with *BEHEMOTH,* it didn't take long to discover the twin effects of a high-tech intriguing contraption that draws a crowd and enough interactivity to convince anyone of dubious motives that they were dealing with something mysteriously intelligent and connected to larger forces Out There. The crowd-pleasing effect makes it impossible for someone to anonymously toss it in the back of a pickup and drive off; people would notice and most likely stand around asking questions. And the interactivity was amazingly effective... when the bike responded to touch by intoning "Do not touch, or you will be vaporized by a laser beam," the effect in some cultures was dramatic, everybody backin' off and lookin' at the bike real s'picious-like. Of course, when it said the same thing one day in Apple's cafeteria, it just caused engineers to swarm all over it while buzzing excitedly about the Berkeley Speech Technologies algorithm and whether they had been detected by microwave Doppler or flux density variations around a puddle of mercury... but I wasn't too worried about those guys haulin' the bike off to a garage and partin' her out.

In a more serious vein, it was comforting to know that the bike would trigger my pocket beeper in response to a security violation, and even do the 911 "I'm being stolen and my coordinates are..." call if the worst happened.

On the Microships, we're taking a similar approach but with more subtlety and bandwidth. If we're nearby and any of a variety of security sensors is triggered, then all sorts of options are available including automatically scanning the area with the video turret and recording or transmitting live footage of the perpetrators, beaconing emergency help requests via packet or satellite, carrying on a running and vaguely threatening commentary with the speech synthesizer, generating loud noises, flashing lights, phoning the Coast Guard or hollering on marine VHF, and unleashing a few nasties. The choice depends on a simple setup tool which allows us to put checkmarks in the intersections of a large matrix of horizontal sensor lines and vertical response lines, then storing various configurations as macros for different environments (marina slips, deserted beaches, campgrounds, urban moorings, friends' garages, and locked corporate meeting rooms all call for different security system configurations).

The nice thing about this, as with most of the other applications, is that we've put so much work into the infrastructure of crossbars and networks that it's trivial to accomplish the rerouting of resources with a few lines of code or mouse clicks on a browser window. The net effect, so to speak, should be a very amusing little pair of boats...

With that, let's turn the clock back eleven years and traverse one last flashback, then wrap this up with a quick look at project management!

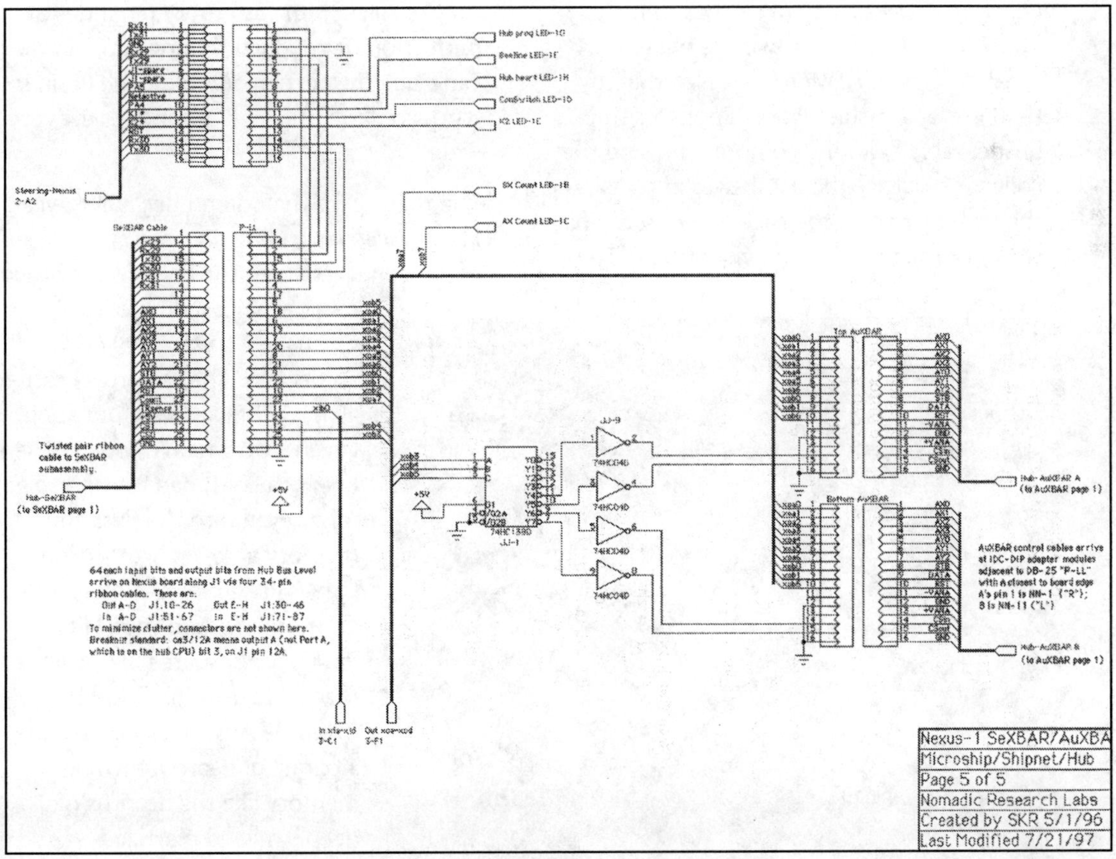

The Microship system is fully documented in the DesignWorks CAD system.

Standing in the Window

May 6, 1989

Somebody on the net recently asked me what my days are like. The popular image, I'm sure, is of a bustling lab with scurrying white-coated technicians fitting glittering surface mount subassemblies onto a frame of high-tech composites and tightly laced bundles of optical fiber and coax. Surrounded by Sun workstations, I coordinate the efforts of teams worldwide, each pouring thousands of man hours into modules of plug-compatible perfection while a software team readies the mega-code that will make it all work. The bike's rollout date is circled on the calendar and we're all feverish, for the eyes of the world are upon us...

Heh. I languish amid the clutter of a household, my back sore from schlepping, my gut swollen from gluttony, my butt blazing red from the Basking Raw-buns syndrome after a day visiting a nudist camp. I whimper, taking in the enormity of the tasks ahead. The *Winnebiko II* is a showpiece hauled to speaking gigs; *BEHEMOTH* is a gleam in my tired eye, an abstract vision. Sponsors ship unspeakably beautiful components, and they collect on particle-board shelves and makeshift tables in my lab... each a reminder of a trick unturned, a bit of magic undone. The TO-DO document, 13 pages and swelling, mocks me: I browse it, looking for a clearly defined task. I seldom find one.

Morning, sunny, a perfect riding day in California. I stare out the window and sigh, for the Other Woman torments me like a curtainless houseful of playful young women next door to a lonely old man. They breeze about in bits of silk, model new bikinis for each other, oil perfect flesh in noonday sun, wake tousled and touch themselves, linger in the perfumed bath, fiercely love their men with cries soft and kisses wet... while he watches, dying inside, desperate moans catching in his throat, the ache as strong as ever but now an agony of frustration. He's obsessed; he can't turn away from the window; frozen dinners burn and telephone solicitors go unanswered when Beauty is afoot. He's in love, in lust, in pain. He knows they're killing him with every reminder of the perfection he touched so long ago, but he can't pull the curtain and hide: the brain is hardwired for SEX, by God, and his eyes are wide and staring.

The road is like that, damn her. For six years I have loved her, riding on wave after wave of passion, every turn the kiss of a new friend, every town a seduction, every campsite a tryst, every downhill an orgasm. The cassettes I carried all have the flavor of "our song"— I even stop what I'm doing when I hear the old road music and gaze misty eyed into the past, my memory an overlay of a dozen scenes linked forever by one musical moment. And the THINGS! Packs, tools, spare parts, toys, even that goddamn shampoo bottle from the Austin hotel... all carry a patina of heartwrenching memory like the detritus of a deeply missed marriage.

So with all this clutter in my head, I surround myself with everything

I fled so long ago—house, furniture, the complex baggage of a life fraught with too many projects and dreams. I stand in the window and watch the road wind sensuously up the mountain, loving other wheels, going on just fine without me, and I'm suddenly that lonely old man, condemned to a life of heartache.

NO!! I recoil, turning enraged from the window. No, no. Not yet, not yet, you bitch, there's still some life in these old wheels. Desperate, I grab the phone and call a potential sponsor, leap onto OrCAD and draw a schematic, wipe clean a section of bench and shove a few parts together. But the energy begins to fade… the project is too big. The monitor array can't be finished because the I/O structure is not defined; that can't be done until the 68000 is running; that won't happen until I build the simulator; that won't work without the monitor array. Aggghhh, hell with it. Another circular problem. Grumbling, I fire up AutoCAD and stare at it, but can't nail down the console layout until I'm sure to have a CMOS VGA LCD driver that can live on the bus and I'm still early in the CAD learning curve anyway… oh, to hell with that too. I furtively play a game of computer solitaire, feebly scan the task list, then wander the house, avoiding the window, plopping at last on the bed to stare at the ceiling and hope the phone will ring to jar me from ennui. (The old man tried and failed; he now stares numbly at the TV with one eye alert to those damn, damn windows next door.)

Maggie wanders in and joins me, eyes full of concern. "Having a bad day, dear?" I try to explain that I am, as Dave Wright pointed out, staggered by my own imagination. But I avoid the central issue and concentrate on the critical-path problems, usually coming around to the fundamental truth that I've bitten off too many projects for one undermotivated tired guy with a bad back. Working on any one of them by definition means I'm ignoring the others, so I do none of them, seeking instead the quick fix, the easy lay, the Clearly Defined Task. She shakes her head, for this is an old story. She snuggles atop me, long fragrant hair flowing into my senses, softness a drug, love a tonic. Time goes away; I murmur sweetness and doze, problems forgotten until later, always later… but then the house fills with housemates and it's dinner, talk, distraction, and a few more stabs at the lab before the evening's drinks fuzz the brain and I give up on one more day.

And as the weeks pass, I watch the calendar, moans catching in my throat. The atlas is a photograph album from an idealized past; I cling to it foolishly and feel a bit embarrassed when Maggie catches me taking it into the bathroom. "Trip planning," I mumble, but we both know the truth: I'm just standing in the window with a sore heart, watching the road go on without me.

Project Management

So. How do we get all this done?

There's a poignancy about it sometimes, this obsessive focus on a mad dream whilst all around me, life continues. Friends age. Babies get old enough to talk, then drive. I wear titanium-framed polycarbonate progressive bifocals; my beard is streaked with gray; I watch employed friends retire on the coattails of the stock options that have kept them chained to their desks for a decade or more. And jeez, it's 2000 already… we've reached the checkpoint of every half-assed article about the future I've ever read. "By the year 2000, we'll all be flying personal helicopters, using video-phones, and living in mile-high buildings made of miracle plastics…"

I was just reading a piece I wrote over a decade ago, early in the throes of the 3.5 year BEHEMOTH project. Same crazy feeling… "standing in the window with a sore heart, watching the road go on without me." Fresh from 16,000 miles of delicious adventure on the *Winnebiko,* I had flung myself into the dream of *BEHEMOTH*… the ultimate expression of technomadic geek bliss. I fondled the Ampro 286 processor board and state of the art 40 MB hard drive with something akin to awe, and having a 68000-0based Macintosh Portable to repackage was on a par with conspicuous consumption, such a luxury was so much power and sophistication in such a small package. But I was almost suffocated by endless details before at last hitting the road again, and by then my fantasies had already leapfrogged my reality: I yearned to move the whole adventure to water.

<creak> So here we are… what is it, two years since leaving Silicon Valley, seven years into the Microship project, 17 since I first pedaled out of Ohio? After all this time, all these labs, all these nautical substrates, and all this work, it's just now beginning to feel real… albeit still vaporous, hypnagogic. I dare not fantasize overmuch about rollicking downriver, and I feel a bit furtive when I sneak the chartbook into the bathroom like a glossy skin magazine to gaze longingly at the pages, one at a time, my mind flowing over seductive benthic topography while I imagine probing estuarine crevasses heavy with the funk of low tide.

It's really all about how we use our time, isn't it? Aging is such a gentle wake up call that you can sleep right through it, watch the Simpsons, have another beer, idly dream of projects, graffitize a passing thought onto the TO-DO list, and call that progress. The back's a little too sore, the shop's a little too cold; what the hell, it'll keep, wouldn't have gotten much done tonight anyway. But then you see an old friend who looks really *old,* and realize with a sudden shock that it's happening to you as well. Damn. How did that happen?

Have you ever calculated how much time you spend watching TV (to pick an easy and obvious example)? A conservative estimate of 1.5 hours per night, year round, amounts over thirteen 40-hour work weeks per year! That's more than one business quarter, or $25,000 of typical Silicon Valley high-tech employee overhead if you prefer to think in Human Resources terms. Add to that the relaxation time, meatspace maintenance, financial management, endless errands, and heaven forbid, the nightmare of commuting… no wonder anyone who gets anything done outside the structured framework of business is almost idolized. I look at these boatlets and cringe at the time wasted enroute… time forever lost in the phosphor vapors of pointless web surfing, the torpor of putzing, the relentless distractions of waiting on hold for customer service idiots and chasing around to fix problems caused by institutional ineptitude.

It's all easy to rationalize, but then, isn't that the whole problem? It takes madness to pull off something like this; it takes crazed obsession, wild-eyed extremism, demonic focus… that hearty roaring piss 'n vinegar of testosterone-infused youth. Natasha looks to me for guidance, and as the Senior Technomad here at NRL I buck up, scan the task lists, issue directives, dive into a project. It's much harder than it used to be, but the Microship, now thrice wet, hovers in the lab like a cruise missile, poised to deliver my explosively restless nomadic spirit to the hotbed of adventure… despite the fact that <red alert> I'm getting dangerously comfortable here.

Volunteers and Geek's Vacations

It's a bit of a challenge, doing something this complex without Deep Pockets and thriving teams of engineers, technicians, machinists, and programmers. We depend heavily on volunteers, obtainium, favors, hand-me-downs, and complete personal dedication to the exclusion of normalcy in all its insidious forms. I periodically update monstrous PERT charts to get a handle on the myriad tasks comprising physical fabrication and system design, each a composite of more than a dozen densely packed laser-printed sheets. We pin these to a big board, stare at them frequently, and color boxes yellow when completed. I make lists, then categorize them and underline the CDT's (Clearly Defined Tasks). The more I accomplish, the longer the lists get… for the more detail I perceive.

Beyond that, it's hard to use formal project management tools, which exist to assign resources to tasks. We're resource limited, so the trick is to just do something whenever possible, periodically going into a productivity frenzy of all nighters in order to make a self-imposed deadline or publicized event.

Our Nomadness mailing list on the Internet (sign up on our website, or email me at wordy@microship.com) has proven to be an amazing resource, giving rise not only to an endless source of ideas and solutions, but also to the ever-popular Geek's Vacation program. This works simply: we invite brilliant tech wizards to come stay with us for a week or two at a time, focusing intensely on some subsystem or fabrication project. In exchange, we put them up in our house, provide food and drink, do a little local sightseeing and ideally share some Time On Water, and make lots of introductions. We've found that the culture that surrounds our technomadic projects is so rich with creative people that this networking is the big draw, and indeed we have seen everything from business partnerships to marriages spin off of first meetings in the bikelab or boatlab.

But mostly we push ahead, year after year, trying to hit a moving technological target without giving in again to *The BEHEMOTH Effect,* lured by the amazing waters just outside our lab and the vivid dream of aquatic technomadness that launched this project one night, so very long ago, on a moonlit lake in the Adirondacks…

Keep an eye on http://www.microship.com. Things are about to get very strange!

MEDIA COVERAGE

Over the years, there have been nearly 600 media appearances of Nomadic Research Labs projects (*Winnebiko I and II, BEHEMOTH,* and the Microships). This has opened doors worldwide, made our sponsors happy, and otherwise augmented all aspects of this adventure. A detailed chronological listing with dates is maintained at http://www.microship.com in the "Press Room" area, but here are some highlights:

Magazine articles

Wired, Sports Illustrated, Bicycling, Cruising World, Cyclist, Discover, Outside, Economist (UK), Futurist, GQ Japan, IEEE Spectrum, Internet World, Dr. Dobbs Journal, Infoworld, MacWeek, Mobile Office, Mondo 2000, Multihulls, National Geographic book series, Newsweek, Omni, People, Popular Computing, Popular Mechanics, Popular Science, QST, CQ VHF, 73, Success, SunWorld Japan, Time, Whole Earth Review, Writer's Digest, Capital (Italy), Asahi (Japan), Telecommute...

Newspaper stories

National

USA Today, Wall Street Journal, Christian Science Monitor, AP wire stories, Stars & Stripes, Star, Newsday...

International

Asahi, London Sunday Times, Independent, Pravda, Guardian, Globe & Irish Independent, Vancouver Sun, Pravda, Australian Women's Weekly...

Local

San Francisco Chronicle, New York Times, Miami Herald, Chicago Tribune, Seattle Times, San Diego Union, San Jose Mercury, Atlanta Journal-Constitution, Boston Herald, and about 200 others...

Television and Radio Shows

National

ABS World News, Guinness Records, CBS Morning News, CNN, Next Step, Nova, Evening Magazine, NBC Earth Journal, NPR All Things Considered, PBS New Tech Times, Synthetic Pleasures film, Phil Donahue Show, New Media News...

International

Beyond 2000 (Aus), Big Byte (Aus), CBC (Canada), BBC Horizon, BBC It'll Never Work, Channel X (UK), TF-1 documentary (France), BBC world service (shortwave), NHK (Japan), Show by Shobai (Japan), TV Tokyo, Vida en Linea, RTE (Ireland)...

Local TV news

Over 150 shows in over 30 states, too many to detail...

PUBLIC SPEAKING ENGAGEMENTS

I never would have believed it back when I considered myself primarily a writer, but the public speaking business has become my main source of income. After doing about 100 book signings and bike presentations around the country back in the late 80's, it suddenly dawned on me that I should be charging for this… and now book engagements both independently and through our speakers bureau (Keynote Speakers, Palo Alto, CA—650.325.8711). If your company or organization is interested in a presentation, possibly with bike and/or boat, please get in touch: this is what keeps the project going forward!

Following are some of the highlights of my speaking gig history; a detailed listing is available on our website at http://www.microship.com.

Corporations

Apple, AT&T, Autodesk, Hewlett-Packard, Solarex, REI, Xanadu, Chips & Technologies, Autodesk, Xerox, IBM, Olivetti, Sun Microsystems, Qualcomm, Bell Labs, Fermilabs, Microsoft, Sharp, National Science Foundation, BBN, Thinking Machines, Motorola, MBARI, Sony, Argonne National Labs, Novell, Northern Telecom, Aspect Telecommunications, Cabletron, Quantum, Life Fitness, Tellabs…

Universities

Stanford, Carnegie-Mellon, University of California (Los Angeles, San Diego and Davis Campuses), MIT, Cal State, University of Houston, University of Washington…

Trade shows or special events

COMDEX, National Online Meeting, Expo '86, West Coast Computer Faire, IHPVA at Smithsonian, American Library Association, EAA Oshkosh, Hackers Conference, SF Great Outdoor Fair, Soviet Silicon Summit, Solar Energy Expo, SolWest, Embedded Systems Conference, WESCON, Seattle Bike Expo, Interop, Sea Kayak Symposium, Mobile World Expo, PacBell Texpo, LA Computer Fair, DECUS, ACM, Usability Professionals Association…

Hamfests

Dayton, Miami, Sarasota, Orlando, Charlotte, Portland, San Jose, Shasta Spacecon, ARRL National and Division conferences, Monterey, Pacificon, Huntsville…

Local events and clubs

Dozens of bicycle clubs, ham clubs, Kiwanis and other organizations, book signings, local computer clubs, professional societies, computer and software user groups, Marine Technology Forum…

MICROSHIP PROJECT SPONSORS

(as of July 2000)

The fundamental design ethic at Nomadic Research Labs is to acquire the best technology solutions available from a wide variety of industries, and then integrate them into a seamless whole. This not only keeps us from committing the sin of wheel-reinvention, but also gives us the luxury of working with some of the most amazing engineers out there. The result is an all-star supporting cast of sponsors and volunteers spanning a wide range of industries. (For information on becoming part of the Microship project through equipment donation, custom work, or direct support, please contact us... we would love to hear from you!)

The following is a list of Microship sponsors; to see the companies that participated in the Winnebiko and BEHEMOTH bicycle projects, please see our website. While there, please take a moment to browse the extensive list of project volunteers. Our odd blend of technology and adventure has attracted some brilliant people, and their creative efforts are reflected throughout these machines.

3Com/Palm Computing
Palm Pilot PDAs and development tools

Addenda Electronics
Serial interface to Sony 8mm VCR

Adept Solutions
Custom Macintosh software

Adobe Systems
PageMaker, IntelliDraw, PageMill, & Illustrator

AFAB
Hydrophone for underwater audio

Alinco
Dual-band handheld and mobile ham radio equipment

Allpen
Custom Newton software & NetHopper

Ampro
PC/104 embedded 286 board and peripherals

Antrim Associates
Marine architecture consulting

Apple Computer
Microship Development Lab 1996-7; Macintoshes

AsepCo
Minolta copier and first lab telephone system

Autodesk
AutoCAD Release 12 for Macintosh

Beehive Technologies
ADB I/O Macintosh interface modules

BiCanoe
SpinFin pedal drive units for Microships

Bicc-Vero Electronics
Speedwire prototyping board

Blue Sea Systems
Marine power wiring products

BoxTop Software
ProJPEG, PhotoGIF, and GIFmation software

Bruninga, Bob, WB4APR
APRSdos tracking software

Calex
DC-DC Converters (triple output)

Capilano Computing Systems, Ltd.
DesignWorks Electronic CAD software

Caribou Mountaineering
Primaloft sleeping bags

Cascade Designs
Dry bags, ThermaRest mattresses, and camp seats

Clippard Instrument Laboratory
Corrosion-resistant hydraulic cylinders

Coherent
Diode laser

CSL
Teflon-impregnated anodizing of rotating mast step

Current Designs
Storm rotomolded kayaks (partial sponsorship)

Dallas Semiconductor
Various integrated circuits & development tools

Dantz
Retrospect backup software for Mac network

Datastick Systems
MyCorder Data Collection interface for Palm

DeLorme Mapping
State atlases

Delrina
Macintosh FAX software

Dimse, Steve
Custom telemetry software

Divinycell (DIAB)
Foam-core material: deck and solar substrate

DraCo Systems
Casablanca nonlinear video editing system

Duofold
Vent-a-Layer clothing

Electric Boat Assoc. of the Americas
Membership, journal, and burgee

Elktronics
Miniature video transmitter kit

eSecure Peripherals
Serial interface to Dallas MicroLAN

Euro Marine Trading
Mobri Radar Reflector

Extrasport
Personal Flotation Device (life jacket)

Microship Project Sponsors

FileMaker/Claris
MacDraw, MacProject, FileMaker, & HyperCard

FineEdge.com
Pacific Northwest nautical books

Fluke
Model 87 Digital multimeter

Formus
CAD work, molds, mechanical consulting

Geoquery
Mapping software with database links

GoLive Systems
CyberStudio web design software

Halted Specialties
Laser printer, test equipment, various electronic parts

Handar
Ultrasonic Wind Sensor (Model 425)

Hewlett-Packard
Laptops, palmtop, calculators

Hewlett-Packard, Elec. Meas. Div.
Logic Dart portable instrument

Hytek Finishes
Anodizing and dichromate sealing of all aluminum parts

Icom America
Amateur Radio Equipment

Infogrip
BAT Chord Keyboard controllers

Interlink
DuraPoint sealed pointing devices

International Medcom
Radalert nuclear radiation monitor

Johnson Camping
Eureka! Equinox 6 dome tent

JWA Marine
Minn-Kota electric thrusters

Kokatat
Water sportswear, foul-weather gear

Konz, Ned
Custom linux server code (in Perl)

Landstrider
Ryan recumbent seats

Lewmar/Simpson-Lawrence
Anchors, mounts, compasses, rudder heads

Magma
Stove/Barbecue

Marinco
Shore power, Nicro vents, and Guest searchlight

Marples, John R.
Marine architecture consulting

Martin, George
Color CCD video camera for turret

Master Appliance
Ultratorch butane soldering tools

Miner Elastomer Products
TecsPak elastomer bumpers for suspension

Mitel Semiconductor
Crosspoint switching chips

Moore Sailboats
Carbon-fiber rudder blades (partial sponsorship)

Morrelli & Melvin
Marine architecture consulting

Motorola
GPS receivers; GMRS, FRS, and marine VHF radios

Mountainsmith
Backpacks and accessories

MPI
Strobe light for PFD

Muench, Bill
FORTH consulting, multitasker, & multidrop code

Napco Security Systems
Security system for development lab

Nelson-Marek Yacht Design
Early marine architecture consulting

Netcetera
Unix sysadmin for www.microship.com

New Micros
68HC11 FORTH boards for control network

Nolan, Tim
Custom solar peak-power tracker and thruster controller

Nuwick
44/120-hour candles

Octagon Systems
Single-board computer for on-board linux server

PacComm
Micropower-2 TNCs for packet radio

PAiA Electronics
Theremax theremin for Europa video control

Panduit
Wiring products

Perimeter Industries
Gunnel Guard integrated fender material

Personal Data Systems
Audapter speech synthesizer

Phoenix Contact
Contact blocks and wiring systems

Power Trends
Integrated switching regulators

Microship Project Sponsors

Qualcomm
Globalstar satellite phone and test account

Quantum
Hard disk drives

Reebok
Hiking boots

Rip and Grip
Hook & loop fastener material

Ritchie
Electronic compass sensor

Roberts, Edward H.
Cash donations

Rocky Mountain Antennas
Dual-band stainless steel J-pole antennas

Ronstan
Blocks and other sailing hardware

Sailrite
Fabrics and Related Materials

San Diego Supercomputer Center
Visualization tools, FEA, early modeling projects

Scripps Institution of Oceanography
Waterproof connectors

Seitech Launching Dollies
Lightweight wheels for landing gear

Shakespeare
5410-XT dual-band cellular antenna

Sharp Electronics
Color active-matrix LCD and micro cameras

Shiva Environmental Systems
Environmental sensor module

Shurflo
Water pumps

Silicon Valley Bus Company
Keystone keyboard/mouse ADB interface

Sky-Pole Manufacturing
Fiberglass tubing

Sobstad Sails
Sail design consulting

Solarex
Photovoltaic modules (1,080 watts)

Sony
D-7 DAT, 8mm VCR, audio amp, speakers

Southwest Windpower
AIR Marine wind generator (partial sponsorship)

Sprague, David & Holly
Cash donation

Sproul, Keith, WU2Z
MacAPRS tracking software

Statpower Technologies
Battery chargers and inverters

Sun Microsystems
SPARCStation, Tek 2465A oscilloscope, office chair

TAPR
MIC-E encoders for APRS

Tonic
Technomad.to domain name in perpetuity

Trace Engineering
C40 charge controller and MicroSine inverter

Traveling Light
Outback oven

Traveling Software
Various LapLink versions

Tyner, Joe
Microship lab telephone system; 386 PC for linux dev.

UC San Diego
Microship lab space 94-95 & student teams

Univenture
CD Viewpack compact-disk storage sleeves

Virtus
Walkthrough Pro modeling software

Warp Speed Imagineering
Macintosh support, wireless networking consultation

Watermark Publishing
Claiborne S. Young's Cruising Guides

Weather Strategies
Dealer for Ocens; Weather seminars

Wenonah
Two Kevlar "Odyssey" canoes for Microship hulls

Werner Paddles
Kayak paddles

Windrider
Two sail rigs (partial sponsorship)

Wireless Networks, Inc.
Wireless Internet link for 97 Microship lab

Yakima
Roof rack for Diesel Mothership

Zocalo
Internet service and web server site

This page unintentionally left blank.

FOR ADDITIONAL INFORMATION

Extensive technical archives on these projects and related resources are available from Nomadic Research Labs. We are slowly producing a family of technical monographs covering all Microship system details, a few key BEHEMOTH components, and various how-to pieces that have spun off of our work on these projects. Please contact us <*wordy@microship.com*> for the most recent information, or write to the address on the back cover for a hardcopy flyer.

Extensive electronic archives are all on our website, of course:

http://www.microship.com

Here you will find the full collection of project status reports, a collection of adventure tales, and much more... updated reasonably often as the project evolves. As we begin traveling full time on the boats, this site will also carry live telemetry and tracking data, images, a clickable breadcrumb trail, and current tales of the adventure.

And finally, we maintain a mailing list on the Net specifically for one-way postings about the project filtered to prevent discussion or noise. To receive updates approximately once a month, subcribe on the website or send us a note with your email address and we'll add you to the nomadness listserv... dont worry, as bona fide Internet old-timers we are vigorously anti-spam and will never use your address for any other purpose!